Learning Mathematics with Micros

ELLIS HORWOOD SERIES IN MATHEMATICS AND ITS APPLICATIONS

Series Editor: Professor G. M. BELL, Chelsea College, University of London
(and within the same series)

Statistics and Operational Research

Editor: B. W. CONOLLY, Chelsea College, University of London

Baldock, G. R. & Bridgeman, T. — **Mathematical Theory of Wave Motion**
de Barra, G. — **Measure Theory and Integration**
Beaumont, G. P. — **Introductory Applied Probability**
Burghes, D. N. & Borrie, M. — **Modelling with Differential Equations**
Burghes, D. N. & Downs, A. M. — **Modern Introduction to Classical Mechanics and Control**
Burghes, D. N. & Graham, A. — **Introduction to Control Theory, including Optimal Control**
Burghes, D. N., Huntley, I. & McDonald, J. — **Applying Mathematics**
Burghes, D. N. & Wood, A. D. — **Mathematical Models in the Social, Management and Life Sciences**
Butkovskiy, A. G. — **Green's Functions and Transfer Functions Handbook**
Butkovskiy, A. G. — **Structure of Distributed Systems**
Chorlton, F. — **Textbook of Dynamics, 2nd Edition**
Chorlton, F. — **Vector and Tensor Methods**
Conolly, B. — **Techniques in Operational Research**
Vol. 1: Queueing Systems
Vol. 2: Models, Search, Randomization
Dunning-Davies, J. — **Mathematical Methods for Mathematicians, Physical Scientists and Engineers**
Eason, G., Coles, C. W., Gettinby, G. — **Mathematics and Statistics for the Bio-sciences**
Exton, H. — **Handbook of Hypergeometric Integrals**
Exton, H. — **Multiple Hypergeometric Functions and Applications**
Faux, I. D. & Pratt, M. J. — **Computational Geometry for Design and Manufacture**
Goodbody, A. M. — **Cartesian Tensors**
Goult, R. J. — **Applied Linear Algebra**
Graham, A. — **Kronecker Products and Matrix Calculus: with Applications**
Graham, A. — **Matrix Theory and Applications for Engineers and Mathematicians**
Griffel, D. H. — **Applied Functional Analysis**
Hoskins, R. F. — **Generalised Functions**
Hunter, S. C. — **Mechanics of Continuous Media, 2nd (Revised) Edition**
Huntley, I. & Johnson, R. M. — **Linear and Nonlinear Differential Equations**
Jaswon, M. A. & Rose, M. A. — **Crystal Symmetry: The Theory of Colour Crystallography**
Jones, A. J. — **Game Theory**
Kemp, K. W. — **Computational Statistics**
Kosinski, W. — **Field Singularities and Wave Analysis in Continuum Mechanics**
Marichev, O. I. — **Integral Transforms of Higher Transcendental Functions**
Meek, B. L. & Fairthorne, S. — **Using Computers**
Muller-Pfeiffer, E. — **Spectral Theory of Ordinary Differential Operators**
Nonweiler, T. R. F. — **Computational Mathematics: An Introduction to Numerical Analysis**
Oliviera-Pinto, F. — **Simulation Concepts in Mathematical Modelling**
Oliviera-Pinto, F. & Conolly, B. W. — **Applicable Mathematics of Non-physical Phenomena**
Rosser, W. G. V. — **An Introduction to Statistical Physics**
Scorer, R. S. — **Environmental Aerodynamics**
Smith, D. K. — **Network Optimisation Practice: A Computational Guide**
Stoodley, K. D. C., Lewis, T. & Stainton, C. L. S. — **Applied Statistical Techniques**
Sweet, M. V. — **Algebra, Geometry and Trigonometry for Science Students**
Temperley, H. N. V. & Trevena, D. H. — **Liquids and Their Properties**
Temperley, H. N. V. — **Graph Theory and Applications**
Twizell, E. H. — **Computational Methods for Partial Differential Equations in Biomedicine**
Wheeler, R. F. — **Rethinking Mathematical Concepts**
Whitehead, J. R. — **The Design and Analysis of Sequential Clinical Trials**

Learning Mathematics with Micros

A. J. OLDKNOW, M.A., M.Tech., F.I.M.A., F.B.C.S.
and D. V. SMITH, B.Sc., M.Sc., Cert.Ed.
Department of Mathematics and Computing
West Sussex Institute of Higher Education
Bognor Regis

ELLIS HORWOOD LIMITED
Publishers · Chichester

Halsted Press: a division of
JOHN WILEY & SONS
New York · Brisbane · Chichester · Ontario

First published in 1983 by
ELLIS HORWOOD LIMITED
Market Cross House, Cooper Street, Chichester, West Sussex, PO19 1EB, England

The publisher's colophon is reproduced from James Gillison's drawing of the ancient Market Cross, Chichester.

Distributors:

Australia, New Zealand, South-east Asia:
Jacaranda-Wiley Ltd., Jacaranda Press,
JOHN WILEY & SONS INC.,
G.P.O. Box 859, Brisbane, Queensland 40001, Australia

Canada:
JOHN WILEY & SONS CANADA LIMITED
22 Worcester Road, Rexdale, Ontario, Canada.

Europe, Africa:
JOHN WILEY & SONS LIMITED
Baffins Lane, Chichester, West Sussex, England.

North and South America and the rest of the world:
Halsted Press: a division of
JOHN WILEY & SONS
605 Third Avenue, New York, N.Y. 10016, U.S.A.

British Library Cataloguing in Publication Data
Oldknow, A. J.
Learning mathematics with micros. –
(Ellis Horwood series in mathematics and its applications)
1. Mathematics – Computer-assisted learning
2. Microprocessors
I. Title II. Smith, D. V.
510'.7'8 QA11

Library of Congress Card No. 83-12642

ISBN 0-85312-513-9 (Ellis Horwood Ltd. – Library Edn.)
ISBN 0-85312-653-4 (Ellis Horwood Ltd. – Student Edn.)
ISBN 0-470-27488-3 (Halsted Press – Library Edn.)
ISBN 0-470-27487-5 (Halsted Press – Student Edn.)

Typeset in Press Roman by Ellis Horwood Ltd.
Printed in Great Britain by Unwin Brothers of Woking.

Table of Contents

Chapter 12 – INTRODUCING CALCULUS

Chapter 13 – NUMERICAL TECHNIQUES

Chapter 14 – APPROXIMATING FUNCTIONS

Chapter 15 – NUMERICAL INVESTIGATIONS

Chapter 16 – PROBLEM SOLVING

Chapter 17 – TRANSFORMATION GEOMETRY

to M^2C

Acknowledgements

The ideas in this book have been tried with teachers, students, colleagues and pupils over the past couple of years and we are grateful to all those concerned with mathematics education who have given us an opportunity to develop work in this area. We have had generous help from bodies such as the Department of Industry, the Department of Education and Science, the Microelectronics Education Programme and the West Sussex Institute of Higher Education in providing hardware and other resources which have been vital in developing our own experience of computer usage in teaching. We are also very appreciative of the continual encouragement and support that we have received from Mike O'Reagan of Research Machines Ltd., Oxford, from Nick Green of Commodore Business Machines, Slough, and David Johnson-Davis and John Coll of Acorn Computers, Cambridge.

Some of the material included here has appeared elsewhere and we are grateful to the Editor of the Association of Teachers of Mathematics publication, *Mathematics Teaching*, for permission to reprint the article concerned with the square shuffle and to Dr. S. H. Hollingdale for permission to use the tables and illustration for the doctor's waiting room simulation from his section on 'Queueing and Simulation' in the Penguin publication, *Newer Uses of Mathematics*, for the Institute of Mathematics and its Applications.

Our thanks go to our colleagues in the Institute's Mathematics Centre for their patient support and advice and, in particular, to Christine Southerton, John Harries and Nigel Bufton for their practical help in producing this book.

Introduction

In the report 'Mathematics Counts' of the Cockcroft committee, a few paragraphs are given to the role of the microprocessor in mathematics teaching. In particular the committee stresses the small amount of work that has yet been done in this field and highlights the need for much more exploration into the ways in which microcomputers may be used for the benefit of mathematics education. We have aimed this book primarily at the teacher of mathematics who has little or no experience of computers or programming but who is interested in exploring the potential of the microcomputer in his own mathematics teaching. We have tried to show that many fruitful mathematical explorations are possible with the aid of very short programs involving quite elementary computing techniques. We have taken the decision to present the programs in BASIC as that remains the programming language most commonly in use across the range of current microcomputers, but all of the programs may be easily converted to other languages such as COMAL or PASCAL.

The programs are written in a style that uses simple statements that have been designed, as far as possible, to run on any machine. The reader may well feel the need to add such refinements as instructions for the use of programs, laying out more carefully the screen output etc. The aim throughout has been to present programs that can be easily understood and adapted by readers who have only an elementary knowledge of programming.

We provide examples of microcomputer applications in many branches of mathematics but we have by no means attempted an exhaustive coverage and we hope the reader will find many areas of his own in which the microcomputer will be of benefit. Again we try not to be dogmatic about how the material should be used – some teachers will want to use some of the techniques presented here in writing their own programs as aids to accompany their mathematics teaching, others may want to use them as problems in their own right for exploration with their pupils or students. Much comment has been made about the lack of 'good software' for mathematics teaching. Comprehensive, easy to use, robust teaching packages are indeed extremely time-consuming to produce, and require many skills other than those of mathematical problem-solving, and our view is that they are best left to the professionals, or to sponsored projects. On the other hand there are many opportunities in mathematics teaching for the use of illustrative programs that may only take, say, ten minutes to produce and which can be prepared outside the classroom or perhaps, developed with the pupils within a normal lesson.

Apart from examples of the use of the microprocessor in teaching the existing curriculum there will inevitably be areas of the curriculum that will change because of the widespread use and availability of computing equipment. At the simplest level there may be changes on notation – perhaps the use of the '*' sign for multiplication, the notation $A(I,J)$ for matrices rather than a_{ij} and so on. Areas of the curriculum may receive different emphases from the present ones – for example it is quite conceivable that the rise of 'computational geometry' may elevate geometry in school to have much greater importance and application than it does currently. The way the machine works may itself bring about changes – perhaps, for example, radians may need to be introduced much earlier as they are used in the SIN, COS, TAN and ATN functions and standard index form may also be encountered much earlier as that is the way calculators and computers represent very large and very small numbers. Again, the availability of computational power may mean that we place greater emphasis on numerical techniques for dealing with general cases, for example Newton–Raphson or Simpson's rule, rather than concentrating on analytic solutions that only apply to a limited subclass of problem, for example the formula for quadratic equations or integration by parts.

Perhaps the greatest potential of the microcomputer lies in its ability to 'speak mathematics'. If we want to control a computer by writing our own programs then, almost without exception, we are involved in applying some mathematics. As more and more households acquire their own microcomputer there is enormous scope for re-awakening an interest in mathematics by our being able to suggest suitably interesting problems for pupils and students to tackle and to help them with the appropriate mathematical techniques. Many thorny issues may spring from this – such as the place of the informal learning of mathematics, whether programming should be considered as a part of mathematics, whether there is a danger of a two-culture division between the computer-lovers and the computer-haters, whether an over-emphasis on computer usage may encourage a self-centred, selfish and introspective attitude in chlidren – but there is no doubt that there are some very exciting times ahead.

1

The anatomy of a microcomputer

1.1 HARDWARE

There are now a huge range of small, portable computers which are just as powerful as the full-scale computers of twenty years ago – the name 'microcomputer' refers, then, rather to their size than their power. At the heart of each microcomputer lies one (or sometimes more) powerful chip called a **microprocessor**. These microprocessors are the major elements in the so-called 'micro-electronic revolution' of the late 1970s and were designed so that they could be programmed to act as controllers for a huge range of devices including washing machines, electronic games, cameras and robots. Although extremely complex the huge volume of sales has meant that the price of these powerful chips has come down so they now retail for just a few pounds.

Without getting too technical, at this point it is worth noting that virtually all the marketed microcomputers at present use just one of two principal microprocessors – the 6502 and the Z80. Microcomputers based on the 6502 include Commodore (PET, VIC), Apple and BBC and on the Z80 include Sinclair (ZX80, ZX81, Spectrum), Tandy and Research Machines.

The microprocessor is controlled by instructions in a language called **machine code** – each microprocessor has its own code. These instructions are usally stored inside other chips and we often think of them as binary numbers although they obviously have an electrical equivalent inside the chips. When a microprocessor is always to obey the same set of instructions (or **program**), such as in a washing machine or hand-held electronic game, the instructions will be held in a special kind of memory chip called a ROM – for Read Only Memory. This kind of memory can best be thought of as permanent and unalterable – the instructions are 'burnt into' it by the manufacturer and cannot be altered in any way once this has happened. Some television games are sold in cartridges that plug into a main unit – in this case the main unit contains the microprocessor and the cartridge contains the ROM into which the instructions for the particular game have been burnt.

With a microcomputer, however, we want, at the least, to be able to type instructions in at the keyboard and to have results displayed on a screen. Some microcomputers (such as the Commodore PET) have the keyboard and screen in the same case as the electronics, others (such as the Research Machines 380Z) have keyboard, screen and electronics all in separate boxes connected with wires. However, the most typical pattern now seems to consist of a case containing electronics and keyboard connected by a wire to the aerial socket of an ordinary TV set. In order for the microcomputer to respond to the keyboard and to send characters to the TV screen the microprocessor must be given a complex set of instructions all the time the microcomputer is switched on. This fundamental program that makes the microprocessor behave like a computer rather than, say, a washing machine is known as its **Operating System** (sometimes abbreviated to O.S.) and it is contained in one or more ROM chips.

If we want to be able to type a fairly lengthy set of instructions in at the keyboard they will each have to be stored inside the computer in another kind of memory chip, called RAM, until we want them to be obeyed. RAM stands for Random Access Memory and this kind of chip can have its contents altered easily, but also needs a constant power supply. As soon as the power supply is switched off (even by a 'blip' on the mains) the contents of a RAM chip are liable to be corrupted and the computer will appear to 'forget' the instructions you typed in.

Thus the essential elements of a microcomputer (called its 'hardware') are a number of chips (microprocessor, ROM and RAM), a means of input (usually a keyboard) and a means of output (usually a signal to a TV set). The chips are normally mounted on a board (called a 'pcb' or 'printed circuit board') on which are printed the electrical connections between components. The board will also usually contain a power supply, some circuitry to 'decode' the keyboard and some circuitry to send out a TV signal (usually a 'character generator' and a UHF modulator). The connections between memory chips and the microprocessor look like several parallel silver stripes and these are known as 'buses' – as they provide the bus-routes (figuratively speaking) along which information moves.

1.2 PROGRAMMING

In order for us to be able to enter instructions to the computer in a language such as BASIC the microprocessor must be given the instructions that enable it to translate each line of the BASIC program into several machine code instructions. The program that does this is called the BASIC interpreter and is usually (with the notable exception of the Research Machines 380Z) provided by the manufacturer in the form of a number of ROM chips fitted inside the machine (sometimes referred to as 'firmware'). Such a program is very long and complex and takes many man-years to develop. Different manufacturers may use different versions of such a program and this leads to variations in which particular features of a supposedly standard language such as BASIC are present in a particular machine. This gives rise to a problem for us since the particular BASIC programs included with this book may need slight modifications before they can work on certain microcomputers.

A line of BASIC program, such as:

```
10 A = B + 2
```

is entered form the keyboard by pressing in the appropriate keys and then pressing the RETURN button to signify that the line is complete. The line of text (which may contain all sorts of programming errors) is stored in RAM character by character. Thus each keyboard character must have its own special code. A typical keyboard may have around 50–60 keys and many keys can be used in conjunction with two special keys: SHIFT and CONTROL. This can give as many as 240 different combinations. If a binary coding is to be used then a single binary digit (or 'bit') can represent one of 2 values (0 or 1), a pair of bits can represent 4 values (00, 01, 10, or 11) and, in general, n bits can represent 2^n values. As $2^8 = 256$ is the nearest power of 2 above 240 we have a collection of 8 bits as an appropriate code for representing keyboard (and screen) characters. Such a block of 8 bits is known as a **byte** and this is the fundamental unit of storage in microcomputers.

Another useful power of 2 is 2^{10}, since $2^{10} = 1024$ which is very nearly one thousand. Hence a collection of 2^{10} bytes is known as a **kilobyte** and usually abbreviated to 1K. The RAM memory capacity of a microcomputer is the one usually advertised when a computer is described as having 8K, 16K or 32K memory. However, not all the available RAM may be available to the BASIC programmer wanting to use the microcomputer.

The operating system and the BASIC interpreter may each need to keep track of the current state of the microcomputer (for example, how much free memory is left) and, as these quantities may change, they cannot themselves be stored in the ROMs. Thus a certain amount of RAM will be used up in keeping records of certain 'housekeeping' variables essential for the operation of the computer. Thus the Commodore PET uses 1K and the BBC micro uses $3\frac{1}{4}$K of the total RAM for its own variables.

Although it does not appear so to us, the picture on a TV screen is redisplayed several times a second and an area of RAM, called the **screen memory**, is needed to keep track of what is currently to be displayed on the screen. If the screen layout consists of 25 lines of text, each 40 characters long, then 1K bytes of RAM are needed for screen memory. If high-resolution graphics are used much more RAM may be needed for the display. Hence, for example, on the BBC model A microcomputer where there is 16K RAM available the operating system needs $3\frac{1}{4}$K for its own variables and a high-resolution graphics mode such as MODE 4 needs 10K for displaying the picture. Hence only $2\frac{3}{4}$K of RAM remains for the BASIC program's use!

In order to save on space, most microcomputers do not save the key words of the BASIC vocabulary such as PRINT, THEN, RETURN as collections of individual characters but use special codes (usually called 'tokens') for each key-word.

1.3 STORAGE OF VARIABLES

When a program is run space will also be required to store the current values of variables used in a program. BASIC has three main kinds of variables:

(i) numeric – usually labelled with a single letter or a letter followed by another letter or digit (although some micros, such as the BBC, allow names of any length) and which may store numbers in either floating-point form or integer form or both, for example:

A BQ C3 ;

(ii) array – labelled with a variable name followed by one or more index numbers in brackets, for example:

V(3) M(5,2)

are elements in arrays;

(iii) string – usually denoted by a variable name followed by a $ sign which can store strings of text characters, for example:

C$ N3$.

Numbers may only be stored to a limited degree of accuracy – see Chapter 10 for more details. Typically each numeric variable may take around 7 bytes of storage, each array variable around 5 bytes and a string will take a little more than as many bytes as it has characters. Thus we can make a crude estimate of the storage requirements of a simple BASIC program, for example:

```
10 DIM N(50)
20 A$ = "THE SQUARE OF": B$ = "IS"
30 FOR I = 1 TO 50
40 N(I) = I*I
50 PRINT A$;I;B$;N(I)
60 NEXT I
```

This program might take around 100 bytes to store. When it is run space will need to be allocated to the numeric variable I, the array variables N(I) and the string variables A$ and B$ and this will take around 300 bytes. Obviously the details will vary between machines, but, in principle, a microcomputer running a BASIC program will have its RAM divided up for use between:

(i) operating system space;
(ii) BASIC program text space;
(iii) variable storage space (numeric, array and string);
(iv) screen memory.

The particular way in which the usage is organised is often illustrated in microcomputer manuals by a diagram called the 'memory map'.

1.4 EXTERNAL STORAGE OF PROGRAMS

Because the contents of RAM will be lost when the computer is switched off, it is essential to have some other kind of storage to which the contents of RAM can be transferred and which will hold them for as long as may be wanted. The most common, and cheap, form of such external storage currently used is audio cassette tape. A microcomputer that uses cassette tape will have instructions within its operating system to LOAD and SAVE programs. When a BASIC program is saved, the contents of RAM containing the program (but not the variables) will be sent out to the cassette recorder as a pattern of low and high sounds corresponding to the bits of the binary numbers representing the bytes of storage. Provided the 'play' and 'record' buttons of the recorder are depressed the audio signals will be copied magnetically onto the surface of the cassette tape. This process of saving does not destroy the contents of the computer's RAM – it just 'reads out aloud' a copy of what is in the RAM to the cassette recorder.

Most computers provide some sort of error-checking process to guard against random 'noise' corrupting the signal sent to the tape recorder; however, it is usually prudent to check that what now appears on the cassette tape is an exact copy of what is contained in RAM. Many microcomputers include an instruction to VERIFY a tape. This entails rewinding the tape and then playing it back into the computer which compares it with the contents of its memory. If there are any discrepancies an error is reported on the screen and you should try to save the program again. Cassette recorders and tape are quite cheap and may not always give perfect results. The tape recorder's heads will need frequent cleaning and demagnetising and it may be wise to avoid the very cheapest tape cassettes.

The **transfer rate** from computer to cassette is measured in a unit called the **baud** which is one bit per second. Typical microcomputer transfer rates are 300 and 1200 baud. A BASIC program that occupies 8K of RAM (quite a long program) would thus take about 4 min to save at the slower speed and 1 min at the faster. Hence there is no need to use long cassette tapes and tapes such as C12 and C15 are quite frequently sold for computer use.

When a program is loaded from cassette the contents of the tape are copied to the RAM of the computer, over-writing what was already there. Obviously it is prudent to keep more than one copy of an important program – preferably storing one copy well away from where the working copies of cassette tapes are stored. Also cassette tapes have little plastic tabs on their cases which can be removed to stop the tape from being erased.

Another important, more convenient, but much more expensive means of storing information is known as a 'floppy disc'. The disc itself is a thin piece of plastic, coated with a magnetic surface and cased in a cardboard cover. The most common discs are about 5 inches across but there are also larger versions (8 inch) and smaller ones (3 inch) in use. The disc itself is not expensive – about £2 each – but it needs to be used with a special piece of electromechanical equipment called a **disc-drive**. This spins the disc rapidly and its magnetic reading and writing heads are able to move close to the surface of the disc. The disc can be thought of as a number (typically 35 or 40) of concentric rings of cassette tape – called **tracks**. As the disc spins the heads can be positioned over any one of the tracks very accurately. Thus much greater transfer rates can be achieved and a typical disc can store around 150K bytes of information. As with cassette tape the microcomputer's operating system will need to include extra commands to handle disc transfers. Sometimes this additional part of the operating system program (called DOS or Disc Operating System) has to be loaded from a disc into the computer's RAM – this process is picturesquely called 'booting' the disc operating system. The same principles of saving, loading and verifying programs apply, but because a disc can hold many programs the need to make 'back-up' copies is even greater. Most discs can be 'write-protected' to avoid them being recorded over by covering a small notch on the casing of the disc with a piece of sticky paper. A disc-drive is a complex piece of precision engineering which can easily be disturbed by misuse, dust etc. It may be wise to budget for maintenance.

1.5 THE VISUAL DISPLAY UNIT (VDU)

The video output from a microcomputer can be in any of a variety of forms particularly where colour output is provided. Essentially there are two main kinds of signal: modulated and unmodulated. A domestic television contains circuitry to decode (demodulate) the signals coming to it from an aerial. In order for a computer to send this kind of signal to the television it must contain a modulator to transform its output, and the television has to be tuned to the particular UHF channel (often 36) that the computer uses. As this process essentially involves two undesirable transforms to the signal, each of which lowers its quality, a much better result is usually obtained by sending an unmodulated signal direct to a monitor – which is like a high-quality TV without the aerial decoding circuitry. In each case the cable is usually a coaxial cable but the plugs on the ends can vary considerably – at the TV or monitor end there is usually either a push-in coaxial TV plug, a screw-in video plug or a bayonet-fit b.n.c. plug. At the computer end the connection may be any of these three types, or a DIN-plug or a jack-plug. It is very useful to have a variety of leads made up for different purposes and it pays to know someone who can solder!

A colour TV or monitor usually has three sperate colour 'guns' – one each for red, green and blue – and a coloured dot on the screen is made by focusing these three beams of colour at a single point with differing intensities. Most sets accept what is called a **composite** colour input. This means that the signal, which can be carried down a single screened cable, consists of separate signals for each of the colour guns sent sequentially. However, the way in which this is achieved varies from country to country. Thus an American microcomputer generating colour signals using the NTSC system is not compatible with a British PAL system colour TV or a French SECAM system colour TV. Some high-class colour monitors use another system, called RGB, in which the signals to the three colour guns are sent in parallel along separate wires.

Monitors are designed so that they can be chained together easily – they usually have sockets for both inputs and outputs, and a switch marked OPEN and TERMINATE (or 75 Ω). To connect a microcomputer with unmodulated video output to two monitors you would connect the microcomputer's output to the input of the first monitor and set its switch to the 'open' position. You would then connect the output of the first monitor to the input of the second monitor and set the second monitor's switch to the 'terminate' position.

To connect the modulated output of a microcomputer to two or more television sets you need a little device called a 'signal splitter', costing about £1. You connect the computer to the input of the splitter and connect each of the outputs to a TV aerial socket. As this process weakens the signal strength it may be necessary to use a rather more sophisticated gadget called a 'splitter–amplifier' which plugs into the mains and costs around £10 and which performs the same task but which boosts the signal when it is split.

It is also possible to feed the video output of a microcomputer into a video recorder. This can be used to produce animated visual materials directly generated by the computer which can be edited and mixed with filmed materials and a sound track to achieve quite elaborate effects. It is also possible to produce OHP transparencies directly from high-contrast photography of the TV output from the computer.

1.6 PRINTERS

The prices of printers have also fallen considerably over the past few years (although not as dramatically as computers) and for some applications, such as school administration, a printer may be almost essential. For program development it is very helpful indeed to be able to obtain a listing of a program that one can take away and work at quietly before returning to develop it further. Essentially there are two kinds of connection between printers and computers ('interfaces') – serial and parallel. In a parallel interface there are, typically, eight wires which simultaneously carry each bit of the binary code representing a printing character. In a serial interface there is just one wire that carries the data to the printer and hence each character is represented by a sequence of, say, eight pulses. Thus a parallel interface is capable of transferring data at a faster rate than a serial one.

There is an international standard system for serial interfaces, called RS232, which is used by many computer peripherals such as printers, plotters and telephone modems. The advantage of this system is that any printer which has an RS232 interface should be able to be used with any microcomputer which provides an RS232 interface. Most printers are mechanical and can receive a lot of wear. It would be wise to budget for maintenance and repairs.

There are, of course, many other devices currently available for microcomputers including input devices such as joysticks, graphics tablets, games 'paddles' and heat, light and sound sensors as well as outputs to make music, speech, control equipment and so on. However, the basic system for a classroom microcomputer need only consist of a microcomputer with keyboard input, cassette storage and TV output. To set up a system with, say, two large second-hand black and white television sets would cost between £200 and £600 at current prices. Obviously if much program development is to be attempted it would be an advantage to have access to a printer and disc-drive (perhaps at a Teachers' Centre or College) but these are expensive items and only become essential when the computer is to be used for administration.

2

Algorithms and procedures

2.1 INTRODUCTION

The word ALGORITHM – like others such as 'feedback', 'interface' etc. – is to be found used liberally throughout the literature nowadays with the implication that we all know what it means. Yet, oddly, our mathematics teaching has been ridden by algorithms for a very long time indeed, though only very rarely has the explicit term 'algorithm' been applied to them. The more common term 'procedure' can usually be substituted for 'algorithm' to make the meaning more accessible.

Most of the alorithms with which we are familiar concern arithmetic. After learning how to manipulate the ten symbols 0, 1, 2, 3, 4, 5, 6, 7, 8, 9 using operations such as addition and multiplication we learn PROCEDURES for long addition, long multiplication and, even, long division. Perhaps a slightly far-fetched notion of a procedure is one which is learnt by practice but which is incredibly awkward to put into words. Just try describing the process of long division without reference to a specific example!

Perhaps a more vivid example comes from the 11–16 algebra syllabus. If you ask someone to write down a formula for solving a QUADRATIC EQUATION it is quite likely that they will admit that they did know a formula once and, perhaps with a little prompting, they might recall something resembling

$$x = \frac{-b \pm \sqrt{(b^2 - 4ac)}}{2a} .$$

A string of symbols buried deep in the memory. In fact it is far more likely that they will be able to recall the formula than remember how to apply it to an equation such as $x^2 - 4x + 3 = 0$.

On the other hand try asking some people to write down the formula for SIMULTANEOUS EQUATIONS. This invariably brings blank looks – perhaps some remember what simultaneous equations are but not how you handle them. Try, then, producing a specific example such as

$$\begin{aligned} 3x + 2y &= 12 \\ 5x - 3y &= 1 \end{aligned}$$

Many people will admit to having met one or other of the PROCEDURES usually called 'elimination' and 'substitution'.

In elimination the aim is to eliminate one of the letters x or y by taking a linear combination of the two equations. Sometimes this procedure is taught quite specifically, for example, 'multiply the first equation through by the x coefficient of the second, multiply the second equation through by the original x coefficient of the first and substract'.

Multiply first by 5: $15x + 10y = 60$
and second by 3: $15x - 9y = 3$
and substract: $19y = 57$

thus $y = 3$, and x is found by substituting this value of y into either of the original equations.

In substitution one of the equations is re-arranged to make either x or y the 'subject of the formula' and that expression is substituted for the corresponding letter in the other equation.

Make y the subject of the first: $y = \dfrac{12 - 3x}{2}$

substitute this in the second: $5x - 3\,\dfrac{12 - 3x}{2} = 1$

and simplify:

$$\begin{aligned} 10x - 36 + 9x &= 2 \\ 19x &= 38 \\ x &= 2 \end{aligned}$$

substitute this in the formula for y: $y = \dfrac{12 - 3 \times 2}{2} = \dfrac{6}{2} = 3$.

In fact there are several other ALGORITHMS for solving 2×2 simultaneous linear equations currently taught of which common ones are 'graphical solution' and 'solution by matrix inversion'. But this proliferation of procedures does not mean that there is no general formula for simulaneous equations. The equations

$$\begin{aligned} ax + by &= c \\ dx + ey &= f \end{aligned}$$

have solutions

$$x = \frac{ce - bf}{ac - bd} \qquad y = \frac{af - cd}{ac - bd}$$

Once upon a time (before calculators and computers) these were taught in the sixth form as part of a course on determinants in the form

$$x = \frac{\begin{vmatrix} c & b \\ f & e \end{vmatrix}}{\begin{vmatrix} a & b \\ d & e \end{vmatrix}} \qquad y = \frac{\begin{vmatrix} a & c \\ d & f \end{vmatrix}}{\begin{vmatrix} a & b \\ d & e \end{vmatrix}}$$

known as Cramer's rule.

Why, then, did someone at some time make the decision that quadratic equations are best solved by using a formula rather than a procedure (such as 'completing the square') whereas simultaneous linear equations are best solved by a procedure rather than a formula (Cramer's rule)? Unfortunately the data on the experiments that this educational time and motions man must have performed on relative efficiency of recall and application are lost to us. Still one might speculate on his wisdom when a whole set of bright sixth-formers all solved the equation $(x - 1)^2 = 9$ by expanding it, rearranging and applying the formula when both sides are already perfect squares.

As with the example of simultaneous equations many problems are amenable to quite a wide variety of algorithms, some of which may be much more efficient than others. In making people think about algorithms and procedures, then, we are constantly concerned with the evaluation of efficiency and our usual starting point comes from a party trick which we call the Dictionary Game.

2.2 THE DICTIONARY GAME

For this game you need a large dictionary, preferably without too many risqué words (such as the 1959 revised edition of *Chambers Twentieth Century Dictionary*), a large blackboard (or equivalent) and a friendly class! Choose a word in advance that you will invite the class to guess. It must appear in the dictionary but, for the greatest success, it should (i) be in fairly common usage, (ii) not be connected with your known personal interests, (iii) have more than 5 letters and (iv) be such that precise answers to questions about its nature can be given. We have found concrete words such as 'museum', 'orphan' and 'piccolo' to be fairly safe bets.

Explain the rules to the class – they will try to guess your word in 20 questions or less to which you will only answer 'yes' or 'no'. To avoid arguments keep a brief record of questions asked and answers given on the board. However, do *not* explain that you are also trying to make it as hard as possible for them to guess the word. Under the guise of appearing to be trying to involve everyone in the game you can ignore questions from anyone who seems to be finding too accurate a line of inquiry and pass to someone else who is throwing up accidental smoke screens!

There seem to be three main kinds of question that are asked:

(i)	semantic	– concentrating on the meaning of the word;
(ii)	grammatical	– trying to identify what kind of word it is;
(iii)	alphanumeric	– trying to find out which letters it has and how many.

A typical set of questions to identify the word 'piccolo' might be:

1	Is it a noun?	Yes
2	Is it animal, vegetable or mineral?	Not allowed (must be Yes/No)
	Is it animal?	No
3	Is it vegetable?	Do you mean 'entirely'?
	Is it partly of wholly vegetable?	Yes
4	Can you eat it?	No
5	Is it mathematical?	No
6	Now many letters does it have?	Not allowed
	Does it have more than 3 letters?	Yes
7	Does it have more than 5 letters?	Yes
8	Does it have more than 6 letters?	Yes
9	Is its first letter a vowel?	No
10	Do you have one at home?	No
11	Have you ever seen one?	Yes, I should think so
12	Is its first letter between 'a' and 'm'?	No
13	Is it made from wood?	Yes
14	Can you sit on it?	Yes
15	Was it designed to be sat on?	No
16	Can you carry it?	Yes
17	Is its first letter between 'n' and 's'?	Yes
18	Is it a tool?	No
19	Is it used in sport?	No
20	Is it 'paintbrush'	No

You then offer the class revenge by getting them to pick a word for you to guess (you might have to stand outside the room while they choose one – hence the need for a friendly class). Suppose they choose 'strontium'. The aim, now, is to show that there is a much more efficient procedure which will home in on any word in the dictionary. So take the dictionary and ask for confirmation that the word is defined in it. Then open it about half way through and choose any word on the open page, for example:

1	Does it appear in the dictionary before the word 'lanyard'?	No

So now keep some fingers of the left hand in the 'lanyard' page and open it half way between there and the end of the book – choosing another word:

2	Does it appear before 'rhubarb'?	No

Move your left hand to the 'rhubarb' page and hunt again between there and the end of the book:

3 Does it appear before the word 'tinder'? Yes

Now the word is sandwiched between 'rhubarb' and 'tinder' so choose a page roughly in the middle of these two:

4 Does it appear before the word 'smock'? No

By this time the procedure should be quite clear – at each stage we will replace one end of the 'sandwich' leaving half the number of words to be eliminated.

5	Does it appear before the word 'strife'?	No
6	'table'?	Yes
7	'sundry'?	Yes
8	'subdue'?	Yes
9	'stuff'?	Yes
10	'strong'?	No
11	'stubble'?	Yes
12	'structure'?	Yes

At this stage there are just 16 words (not counting derivations) in *Chambers* between 'strong' (inclusive) and 'structure' (exclusive) – so we are bound to get there within the 8 questions left if we stick to this procedure.

13	Does it appear before the word 'strophe'?	Yes
14	'strontium'?	No
15	'strop'?	Yes

Now there are just two choices, so a definite question can be used – as the choices are 'strontium', known from school chemistry, and 'strook', an obscure word used by Spenser, Shakespeare and Milton, a little deduction may be used.

16 Is the word 'strontium'? Yes

We have taken the trouble to labour this example because we have found that it gives a memorable introduction for students to the power of a search procedure that can be applied in a variety of situations and which is sometimes called a 'binary chop'. At this point it is worth raising the question, 'How many words are there in the dictionary?'

With one question you can decide between two alternatives, with two questions that becomes four and so on.

No. of questions	No. of words
1	2
2	4
3	8
4	16
5	32
⋮	⋮

In our example we took 16 questions so we have an estimate of 2^{16} words in the dictionary. This forms a reasonable point to get to grips with powers of 2 – in particular with 2^{10}. Now $2^{10} = 1024$ which is quite close to a thousand or 10^3. In the computer world 2^{10} is referred to as 1K, where K stands for kilo. Thus a microcomputer described as having 16K of memory actually has $16 \times 1024 = 16\,384$ bytes of memory – the approximation that 1K is a thousand gives an underestimate of around 2%. Some people even give their salaries nowadays in kiloquid! Thus a salary of £8500 would be written as £8.5K. From the approximation that $2^{10} \approx 10^3$ we have that $2 \approx 10^{0.3}$ so that the log of 2 to the base 10 is about 0.3 (compared with the value of 0.3010 from 4-figure tables).

All this, though, is a slight diversion as the aim was to feel 'at home' with the size of 2^{16}. Now $2^{16} = 2^{6+10} = 2^6 \times 2^{10} = (2^3)^2 \times 2^{10} = 8^2 \times 2^{10} = 64\text{K}$ which is a little more than 64 000.

In fact, though, the numbers in the 'No. of words' column are all maxima. We can interpret the table in a slightly different way by saying that, for example, any number of words between 17 and 32 can be eliminated in 5 questions. Thus our value of 16 questions for the dictionary would give an estimate that lay between $2^{15} + 1$ and 2^{16} words, that is between about 32 000 and 64 000 words. There is also scope for interesting discussions to be had about the unevenness of the distribution of words between pages of the dictionary and about the question of luck – there is room for much interesting statistical research into the validity of our crude assumptions.

Now that we have an example of a problem which can be tackled by a number of algorithms – one which is markedly more efficient than the others – we turn to the problem of notating algorithms. As was previously mentioned we usually learn algorithms by following worked examples from which we each try to induce a set of general rules in our kind of 'mind language'. The validity of our own internalised rules can then be tested by our performing many drill examples to search for any errors that may be symptoms of our 'confusion'. Only very rarely indeed are a set of such general rules written out in natural language. When they are, the appearance is usually so clumsy as to make them of no practical use. A close analogy can be found in the way we learn games such as cribbage, whist, chess etc. Here the rules, together with their interpretation, are usually best learnt by playing with patient partners who will explain problems as they arise, rather than by studying a book on the subject.

2.3 RUSSIAN PEASANT MULTIPLICATION

An example we usually use for this purpose is the so-called Russian Peasant Multiplication Algorithm which we attempt to teach by worked examples on the board without any words being spoken at all. After a few minutes the board is covered with examples such as:

29	17		12	35		19	18
58	8		6	70		38	9
116	4		3	140		76	4
232	2		1	280		152	2
464	1					304	1
493				420		342	

It is easy to see that one column of figures is being doubled while the other is halved (with remainders omitted) and that the result is each case is given by multiplying the numbers at the heads of the columns. It is also fairly easy to see that the result is also the sum of the underlined numbers in the colum which is being doubled. However, it is not so easy to see how the underlined numbers were singled out from the rest – is it just a coincidence that each example contains just two underlined numbers or that the last figure to be doubled is also underlined? Some more examples may serve to dispel or reinforce such hypotheses.

23	16
46	8
92	4
184	2
368	1
368	

15	18
7	36
3	72
1	144
	270

It can take quite some time before the suggestion is made that the underlined numbers correspond to the *odd* numbers in the halving column. After a little while most people are able to get the hang of this algorithm and to apply it successfully to examples of their own – checking by performing the long multiplication. In fact it is often found so simple that people wonder why they were not taught it instead of long multiplication. Some people may also be able to see *why* the algorithm works.

However, the purpose of this exercise is not to teach a new way of multiplying but to motivate the need for a system of notation for algorithms, so we usually ask for volunteers to explain in words how the algorithm is formed. This can, of course, lead to a variety of more or less coherent forms. A possible attempt might run as follows: 'Start with the smaller number. If it is odd add the larger number to a running total. Continually halve the smaller number, ignoring any remainder, and double the larger, adding the larger to the total if the smaller is odd until the smaller reaches 1. Add the final larger number to the total and the total now holds the answer'.

This attempt to specify an algorithm in natural language certainly captures the spirit of the process and, although slightly clumsy, is largely unambiguous. This particular algorithm is one that involves the same set of operations being performed over and over again, terminated by some tested condition. Very many algorithms are of this type and the repeated instructions are often known as a 'tested loop'.

2.4 NOTATING ALGORITHMS

For the exchange of algorithms between practising mathematicians, computer scientists etc. a rather more formal language is desirable, yet no commonly agreed standard exists. Those attempts that have been made all use a form of 'transformation algebra' which denotes the replacement of the value of a symbol by a well-formed expression. A simple such form is used by Knuth. In this system the algorithm might take the form:

ALGORITHM R (Russian Peasant Multiplication) given two positive integers M and N to find their product T.

R1 (Find the smaller) Set $S \leftarrow M$, $L \leftarrow N$, $T \leftarrow 0$. If $M > N$ then set $S \leftarrow N$, $L \leftarrow M$. (Now S is the smaller number and L is the larger.)
R2 (Is S odd?) If S is odd then set $T \leftarrow T + L$.
R3 (Have we finished?) If $S = 1$ then T holds the product and the algorithm terminates otherwise
R4 (Halve and double) Set $S \leftarrow \lfloor S/2 \rfloor$, $L \leftarrow 2L$. Go back to step R2.

The function $\lfloor x \rfloor$ is called the 'floor' of x and means the largest integer smaller than or equal to x. Here each step of the algorithm is numbered and begins with a brief phrase to sum up the principal contents of the step. Thus the algorithm can be accompanied by a simple FLOWCHART to show more clearly the way in which the algorithm works.

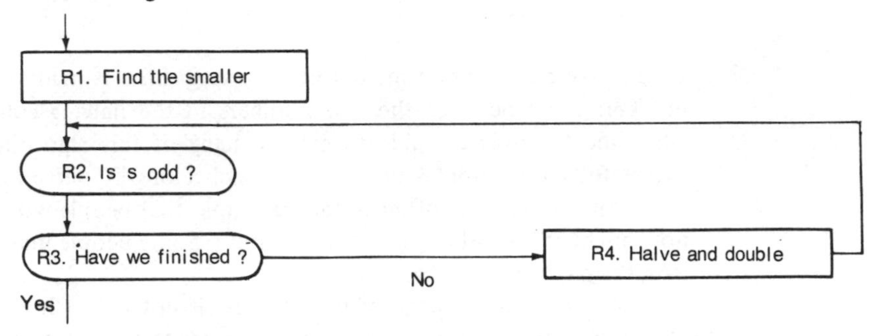

This mixture of algebraic notation and explanation in natural language is quite a simple form both to understand and to produce. The notation '$x \leftarrow y$' means 'the variable x now has the value which is currently the value of y' and it does not in any way alter the value held in y. A notation such as '$n \leftarrow n + 1$' is often read as 'n is replaced by $n + 1$' and means that the value of the variable n is replaced by one more than its present value.

Over recent years, though, several mathematics syllabuses have required the teaching of a more or less standardised system for flowcharts. Here there is a convention on the shape of 'boxes' to surround steps of the algorithm:

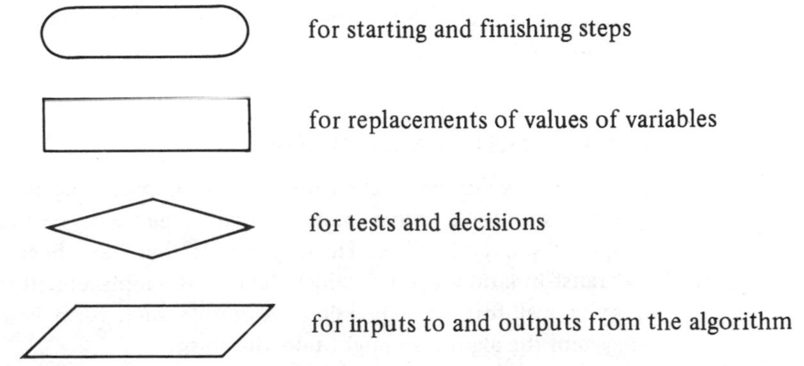

However, there is less standardisation about the notation for the replacement operation. Some texts use '$x \leftarrow y$', as above, others use '$x := y$', but there does seem to be general agreement that the destination variable is named on the *left* of the replacement sign. In this convention our algorithm might appear as shown in Fig. 2.1. Here the floor function notation $[x]$ has been replaced by another notation INT(x). There has been considerable debate recently about the value of this type of flowchart. Although there is an almost standard convention for flowcharts they are tedious to produce, can be very sprawling and difficult to follow and contain no helpful annotations in natural language. Their main advantage seems to be that they model quite closely the way in which a computer program may be written in certain programming languages such as FORTRAN or BASIC.

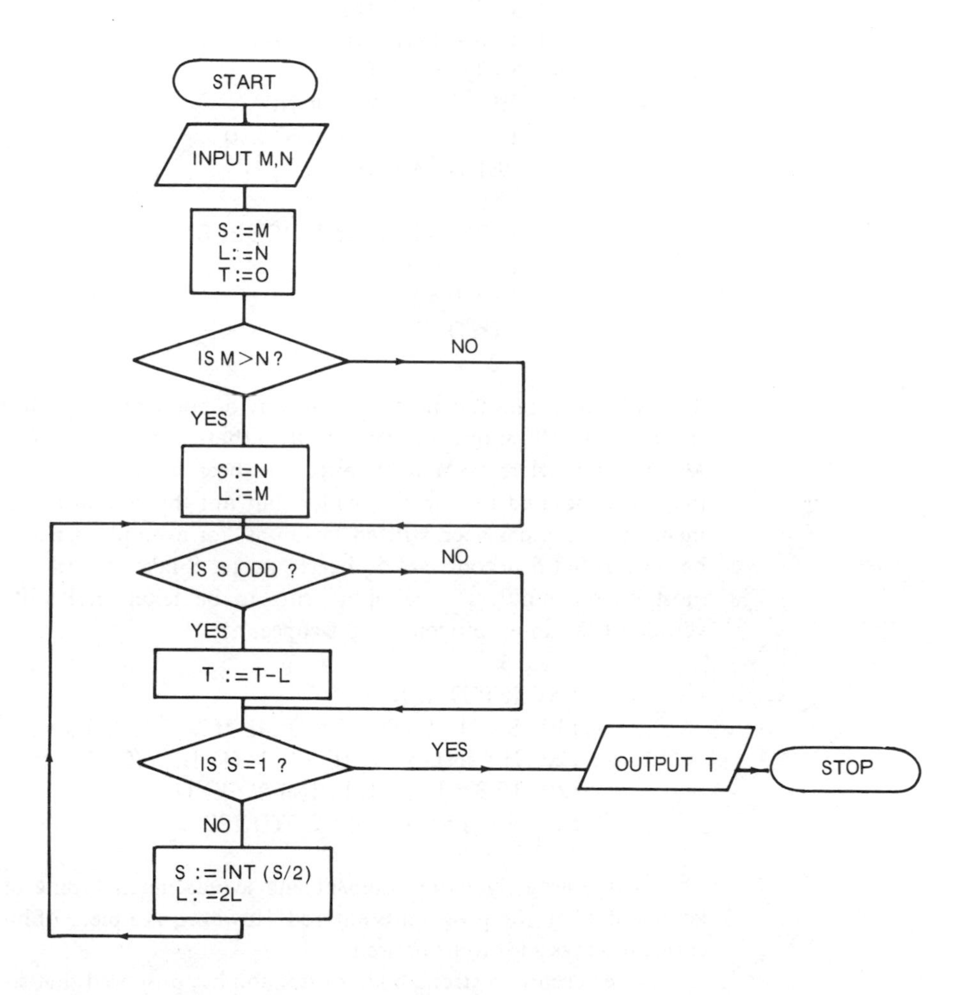

Fig. 2.1

It might seem, then, quite a reasonable step to use a computer programming language as the means of defining an algorithm. Thus many journals now publish algorithms in the form of computer programs. For example a BASIC equivalent of the flowchart might be:

```
 10 REM RUSSIAN PEASANT MULTIPLICATION
100 INPUT M, N
105 REM FIND THE SMALLER
110 LET S = M
120 LET L = N
130 LET T = O
140 IF M < N THEN GOTO 170
150 LET S = N
160 LET L = M
165 REM IS S ODD?
170 LET H = INT(S/2)
180 IF S = 2*H THEN GOTO 200
190 LET T = T + L
195 REM HAVE WE DONE?
200 IF S > 1 THEN GOTO 230
210 PRINT "ANSWER IS"; T
220 STOP
225 REM HALVE AND DOUBLE
230 LET S = H
240 LET L = L*2
250 GOTO 170
```

This is by no means the most efficient way of coding the algorithm (for example if the test in 180 shows S to be even then there is no need to goto to line 200 to test if $S = 1$). Here REM statements have been used to show how the program may be annotated to help the reader (but not the computer!) and the replacement operation has been written in full as, for example, LET S = M which can be read as 'let S become M' or, better, as 'let S take the value of M'. However, most microcomputers allow short cuts to be taken in BASIC. A condensed version of the above program might appear as:

```
100 INPUT M, N
110 S = M : L = N : T = O : IF M > N THEN S = N : L = M
120 H = INT(S/2) : IF S > 2*H THEN T = T + L
130 IF S = 1 THEN PRINT T : END
140 S = H : L = 2*L : GOTO 120
```

This is undoubtedly more compact and avoids much typing at the computer keyboard when the program is entered. However, as a piece of human communication it leaves a lot to be desired.

More recently a strong body of opinion has proposed that so-called STRUCTURED programming languages such as PASCAL and COMAL bridge the gaps between algorithm design, algorithm interchange between humans and writing programs for computers. A COMAL version of the Russian Peasant Algorithm might be:

```
INPUT first, second
IF first < second THEN
   small := first
   large := second
ELSE
   small := second
   large := first
ENDIF
half := INT(small/2)
IF small = 2*half THEN
   total := 0
ELSE
   total := large
ENDIF
WHILE small > 1 DO
   small := half
   large := 2*large
   half := INT(small/2)
   IF small > 2*half THEN
      total := total + large
   ENDIF
ENDWHILE
PRINT "Answer is"; total
END
```

There is also an associated convention for a graphic representation, called a STRUCTURE DIAGRAM, to illustrate COMAL representation of algorithms (see Atherton [2] for more detail). Undoubtedly this convention for writing and illustrating algorithms has much to commend it. It is also quite easy to translate most COMAL and PASCAL progams into (usually longer) BASIC ones. However, as the vast majority of the current generation of microcomputers use BASIC as their standard language and as there is, as yet, no widely available COMAL interpreter for microcomputers we have decided to present algorithms in this book in a version of BASIC that should be easy to enter into most common microcomputers and should also be fairly straightforward for humans to follow. Some versions of BASIC do include 'structured statements' such as IF ... THEN ... ELSE ... and REPEAT ... UNTIL ..., as well as having long names for variables that allow both upper and lower case symbols. Again, though, we have chosen to use something like a 'lowest common denominator' for BASIC in the interests of widespread applicability. Fortunately the algorithms used in this book are quite simple and we hope that purists will not be offended by our decisions.

In summary, then, we are firmly convinced that the production of algorithms, together with their testing for efficiency, is an important mathematical activity that could be reflected in the curriculum. The means of expressing algorithms is a personal choice although it will inevitably benefit from the applicaton of ideas from logic and algebra. To this end the inclusion of 'flowcharting' in mathematics syllabuses is a well-intentioned development along these lines. However, this is often counter-productive because of the tedium of accurate drawing and because its conventions are not wide enough to encompass the structures useful in designing algorithms. A programming language itself is usually a more acceptable medium with the enormous advantage that, with a microcomputer available in the mathematics class, algorithms can be immediately tested. We feel strongly, then, that programming has a vitally important part to play in mathematical education and that pupils should have, wherever practicable, an opportunity to apply their mathematics using a computer.

Returning to the Russian Peasant Algorithm remember that we have not proved that it works nor explained why it works (which might amount to the same thing) but just found some ways of describing HOW it works. Before moving to those chapters which describe some algorithms the reader is invited to tackle a problem of algorithm design that will be taken up in a later chapter.

2.5 A MAGIC SQUARE PROBLEM

In a 4 by 4 magic square each row, column and diagonal uses four difficult integers between 1 and 16 and its total is the 'magic number' 34. How many different choices of 4 different integers between 1 and 16 have such a total? If we write out sets of 4 numbers is ascending order then possible choices are (7, 8, 9, 10), (3, 7, 9, 15), (1, 2, 15, 16) etc. We seek a systematic way of scanning through the possible sets of 4 different numbers to find which have a total of 34. A simple approach is illustrated by the following program:

```
100 REM A SEARCH FOR 4X4 MAGIC SQUARE TOTALS
110 LET S = 0
120 FOR I = 1 TO 13
130    FOR J = I + 1 TO 14
140       FOR K = J + 1 TO 15
150          FOR L = K + 1 TO 16
160             IF I + J + K + L = 34 THEN LET S = S + 1
170          NEXT L
180       NEXT K
190    NEXT J
200 NEXT I
210 PRINT S
```

This program makes good use of so-called 'nested loops' which are highlighted by the indentations. Try running the program and see how long it takes to produce the result – which should be 86.

It should be quite easy, then, to extend this problem to find similar totals for 6 × 6, 8 × 8, 10 × 10 magic squares. For the 6 × 6 square the numbers are taken 6 at a time from the 36 numbers between 1 and 36 and their total is

$((1 + 36)/2) \times 6 = 111$. Only minor changes will be needed to the previous program – the ranges of the I, J, K and L loops will change, two more nested loops (M and N say) will be needed and the test on line 160 will become:

```
IF I + J + K + L + M + N = 111 THEN LET S = S + 1 .
```

Try running the program, but do not be too surprised if it runs for a long time without printing anything. At any stage you should be able to interrupt the program using the STOP, ESCAPE or control code keys that your particular microcomputer recognises as the signal to break into a program. You can then ask to have the current values of the variables printed out by giving a direct instruction (that is, with *no* line number):

```
PRINT I, J, K, L, M, N, S .
```

With most microcomputers you can resume the program from where you interrupted it by typing CONT (for 'continue') – if you type RUN the program will start again right from the beginning. Can you find a more efficient procedure that will produce the required result in a much shorter time?

3

Functions and equations

3.1 FUNCTIONS

A very simple, and yet both powerful and general, program to produce is one which will tabulate the values of a given function over a range of values.

Program 3.1 – To tabulate a function

```
100  REM PROGRAM 3.1 – TO TABULATE A FUNCTION
110  DEF FNA(X) = X*X - 4*X + 2
120  XL = -10 : XU = 10 : XS = 1
130  PRINT "X", "Y"
140  FOR X = XL TO XU STEP XS
150     PRINT X, FNA(X)
160  NEXT X
```

The shape of the graph of a function can be simply induced from the pattern of numbers in the Y column and also the position of any roots in the range of values for X may be roughly located. By changing the values in line 120 (perhaps by using an INPUT statement) we can 'zoom-in' on particular parts of the range of the function. Thus changing line 120 to

```
120  XL = 3 : XU = 4 : XS = 0.1
```

we can hunt for the root of the quadratic equation $x^2 - 4x + 2 = 0$ that lies between $x = 3$ and $x = 4$. Similarly by substituting

```
120  XL = 3.4 : XU = 3.5 : XS = 0.01
```

it is clear that a root lies between $x = 3.41$ and $x = 3.42$.

This technique of 'zooming-in' can also be very useful in hunting for maxima and minima of functions. It can also be used to find solutions of non-linear simultaneous equations and to study functions of several variables. As an example we shall apply the technique to a problem which tackled conventionally leads to a very awkward equation to solve.

3.2 THE BOOKSHELF PROBLEM

A long bookcase, one foot wide, is to be moved down a long corridor 4 feet wide and 7 feet high. What is the height of the tallest bookcase that can pass down the corridor?

Suppose the bookcase is h feet tall and is inclined to the horizontal at an angle θ, as in Fig. 3.1. By considering the length and breadth of the corridor we can derive the equations:

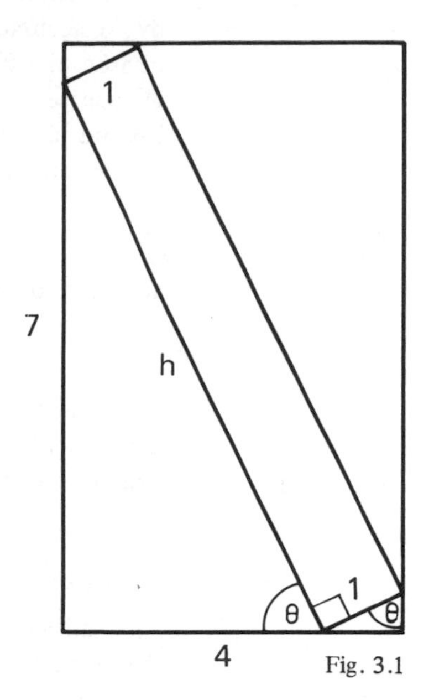

Fig. 3.1

$$h \cos\theta + \sin\theta = 4 \tag{3.1}$$

$$h \sin\theta + \cos\theta = 7 \tag{3.2}$$

Re-arranging these gives the pair of equations $h = (4 - \sin\theta)/\cos\theta$ and $h = (7 - \cos\theta)/\sin\theta$ and we seek a solution for h (>7) corresponding to a value of θ between 0° and 90°.

Program 3.2 – To tabulate a pair of functions

```
100  REM PROGRAM 3.2 – TO TABULATE A PAIR OF
     FUNCTIONS
110  DEF FNA(T) = (4 - SIN(T))/COS(T)
120  DEF FNB(T) = (7 - COS(T))/SIN(T)
130  TL = 10 : TU = 80 : TS = 5 : DR = 3.1415927/180
140  FOR TD = TL TO TU STEP TS
150     T = TD*DR
160     PRINT TD, FNA(T), FNB(T)
170  NEXT TD
```

In Program 3.2 (above), TD is the angle θ in degrees and T is its radian equivalent. Obviously we must take care not to let either SIN(T) or COS(T) become zero and hence the limits, TL and TU, of the range of values for TD. A first run of the program reveals an intersection of the graphs of the two functions in the interval $60° < TD < 70°$. Running the program again with TL = 60: TU = 70: TS = 1 gives a value of h about 7.3 for TD between 64° and 65°. By continuing to zoom-in we arrive at a value for h of about 7.266 with TD just above 64.787°.

Of course there is a limit to how far we can continue this process determined by the accuracy to which the computer can handle numbers. Typically this would be around 7 significant figures, but see Chapter 10 for a more detailed discussion. Of course for many 'real-world' problems there would be no point in going beyond, say, 4 or 5 significant figures.

We could, of course, have eliminated h between (3.1) and (3.2) to give

$$4 \sin\theta - \sin^2\theta = 7\cos\theta - \cos^2\theta$$

and then to have hunted for a root of the equation $f(\theta) = 0$ where

$$f(\theta) = (4 - \sin\theta)\sin\theta - (7 - \cos\theta)\cos\theta$$

In order to produce a direct solution for h we would have needed to eliminate θ between (3.1) and (3.2) which is possible, but awkward, and yields a quartic equation for h.

3.3 COPS AND ROBBERS

An example of a problem involving a function of two variables comes from the theory of games.

A security firm regularly carries a company's payroll from a bank to a factory. The van carrying the money can take one of two possible routes – either through the countryside. A gang of robbers want to ambush the van. If they ambush the van in the town there is a probability of 0.7 that the attack will be successful. If they ambush it in the countryside the probability of a successful attack is 0.6. There are only enough members of the gang to set up an ambush on just one of the two routes at each occasion. What strategy should the security firm adopt in choosing the routes to maximise their chances of getting the payroll through? What strategy should the robbers adopt about their choice of route to maximise their chances of a successful attack?

Obviously neither the security firm nor the robbers should adopt a completely regular pattern of routes. They need to make a random choice on each occasion but the means of selecting the choice must reflect the most advantageous mix of the two routes. Suppose the security firm select the probability x with which the van will take the town route. The robbers select, independently, the probability y that they will ambush the town route. From this information we can build up a contingency table of the probabilities of the van getting through.

The top left-hand entry in Table 3.1 corresponds to both the robbers and the firm choosing the town route on a given day – which happens with probability xy. The table then shows that there is a probability of 0.3 that the resulting attack is unsuccessful. Hence the probability of both the robbers and the firm choosing the town route and the payroll getting through in $0.3xy$. Similarly the top right-hand entry in Table 3.1 shows that if the robbers choose the town and the firm chooses the countryside then the payroll is certain to get through. This choice of routes has probability $(1-x)y$ and hence the probability that the van gets through with this pattern of routes is $1.(1-x)y$. If we introduce the function $P(x,y)$ as the probability that the van gets through for given values of x and y then, by summing the probabilities for each of the four possible combinations of route, we have

$$\begin{aligned} P(x,y) &= 0.3xy + 1.(1-x)y + 1.x(1-y) + 0.4(1-x)(1-y) \\ &= 0.4 + 0.6y + 0.6x - 1.3xy \end{aligned}$$

The robbers want to choose the value of y that minimises this function and the firm want to choose their value of x to maximise it. In order to investigate the behaviour of the function we can choose to fix a value of x, say, and tabulate $P(x,y)$ for a suitable set of values of y. By doing this for each of a suitable set of values for x we have a series of tables that correspond to a set of parallel 'slices' through a surface in three dimensions. By tabulating values of x corresponding to given values of y we can obtain information about slices through the surface perpendicular to the previous ones. Since x and y are both probabilities we are only interested in values of $P(x,y)$ for $0 \leqslant x, y \leqslant 1$.

Table 3.1

			Firm	
			Town x	Country $1-x$
Robbers	Town	y	0.3	1
	Country	$1-y$	1	0.4

Program 3.3 – To tabulate a function of two variables

```
100 REM PROGRAM 3.3 – TO TABULATE A FUNCTION OF
        TWO VARIABLES
110 XL = 0 : XU = 1 : XS = 0.1
120 YL = 0 : YU = 1 : YS = 0.1
130 FOR X = XL TO XU STEP XS
140    FOR Y = YL TO YU STEP YS
150       P = 0.4 + 0.6*Y + 0.6*X - 1.3*X*Y
160       PRINT X, Y, P
170    NEXT Y
180    PRINT "PRESS A KEY"; : INPUT A$
190    PRINT
200 NEXT X
```

Line 180 of Program 3.3 (above) is just to allow a delay between each of the tables appearing. By observing the gaps between the value of P in each of the tables we can see that, for fixed x, P is a linear function of y. Also the 'slope' of each of the lines becomes smaller as x increases. For $x = 0.4$ the increments between values of P are 0.008 and for $x = 0.5$ they are -0.005. This suggests that there is a value of x between 0.4 and 0.5 for which the function $P(x,y)$ is constant – that is, independent of y. This value can easily be sought by amending line 110 to 110 XL = 0.4 : XU = 0.5 : XS = 0.01. From this we find that the value of x lies between 0.46 and 0.47 and a further 'zoom' shows that x lies between 0.461 and 0.462 and gives a constant value for P of about 0.677.

By reversing the order of the X and Y loops we can perform exactly the same investigation into the behaviour of $P(x,y)$ as y changes. Because of the symmetry of the form of the function $P(x,y)$ it is clear that there is a value of y between 0.46 and 0.47 for which $P(x,y)$ is independent of x and at which it has a constant value of about 0.677. We can re-arrange the form of $P(x,y)$ to give

$$P(x,y) = (0.4 + 0.6x) + y(0.6 - 1.3x)$$

which shows that P is indeed linear in y for fixed x. The slope of the line is zero when $0.6 = 1.3x$ which yields $x = 0.6/1.3 \approx 0.4615$. Substituting in $P(x,y)$:

$$P\left(\frac{0.6}{1.3}, y\right) = 0.4 + 0.6\,\frac{0.6}{1.3} = \frac{0.88}{1.30} \approx 0.6769\ .$$

Although the algebraic analysis is obvious to an experienced mathematician many learners will find that the empirical evidence given by such a program provides the necessary 'clue' (and motivation) to start a search for an algebraic explanation. The process is quite similar to a scientifc experiment in which we hold all other parameters steady while observing the behaviour as just one parameter is varied. From these observations we can make some hypotheses and that leads to a search for the theory to substantiate or reject them.

3.4 THE SOLUTION OF ALGEBRAIC EQUATIONS

In a common approach to equations questions are posed such as 'Find the value of x for which $3x = 24$?' or '$x + 7 = 12$'. The questions may be answered either by simple arithmetical processes or by trial and error. For many pupils it is the latter that is the natural approach. As the topic develops manipulative techniques are learnt and it is often these that are expected and tested in later work. The spontaneous idea of trying some numbers to find one that fits is often actively discouraged. If the equation arises from some problem situation, as indeed to maintain realism it often should, then the situation itself is one of 'find the number'. Further the over-emphasis on stock manipulative methods leads to a restriction on problems that might be tackled to those that lead to linear or quadratic or other simple equations. Though we do not in any way wish to deprecate the value of the development of manipulative skills, we do believe that methods developed from a trial and error approach are equally valuable, and with the aid of a calculator or computer can lead to the development of some sound numerical work. There is also no limit on the type of equation that may be successfully tackled even in elementary work. The early part of this chapter has shown how a step by step function evaluation program may be applied to solve equations. We will now look at some more problem situations to illustrate how a trial and error approach may be developed into a sound method of solution.

3.5 THE OPEN BOX PROBLEM

Cardboard is available in rectangular sheets measuring 50cm by 60cm. A square is cut from each corner and the sides folded up to make an open box with a volume of 9000 cm^3. What size should the squares be?

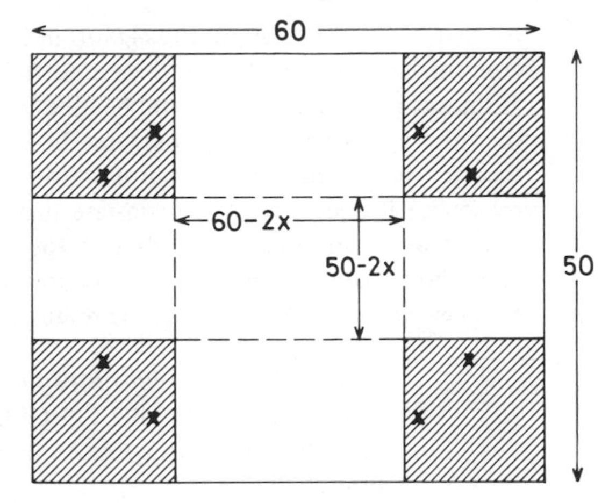

A natural way to solve this problem is to take a value for x and calculate the resulting volume. Several values can quickly be tried and the numbers involved organised into a table:

x = box depth	width	length	volume	
10	30	40	12000	Too large
8	34	44	11968	Better, but still too large
6	38	48	10944	Too large, but better still
4	42	52	8736	Too small.

At this point we know that there is a solution to the problem in the interval [4, 6].

Before going on to refine the answer it is useful to state what we are doing algebraically. The height of the resulting box is x, the length is $60-2x$ and the width is $50-2x$. Hence

$$\text{volume} = x(50-2x)(60-2x)$$

so that the search is for a solution of the equation

$$x(50-2x)(60-2x) = 9000 \tag{3.3}$$

It does not take very long with a calculator to find a solution to (3.3) by trial and error but if a computer is available then the process is even faster. First we need to develop a sensible strategy and this may be put in the form of an algorithm.

Algorithm for solving the open box problem

(1) Find two values of x such that for XS the volume is too small and for XL it is too large.
(2) Calculate XM = (XS + XL)/2.
(3) Calculate the volume V(XM).
(4) If V(XM) is too small then put XS = XM else put XL = XM.
(5) If XL − XS is not yet small enough then go back to (2).
(6) Output 'The required value lies in the interval' [XS, XL].

All that is needed to apply the algorithm is a simple computer program such as Program 3.4 (below). On a microcomputer it is not necessary to provide an end in the program, the stop key can be used to halt execution. Those who have not tried the approach before should see how long it takes to get, say, six figure accuracy in the answer. Of course we do not usually require such an accurate result, but it does serve to demonstate that a method developed from trial and error is a realistic one and leads to a sound numerical method. The algorithm given above may be turned into a computer program in which the computer tests each new value of x and outputs the result when the interval is small enough.

Program 3.4 – A simple program for the open box problem

```
100 REM**OPEN BOX PROBLEM**
110 PRINT "WHAT VALUE FOR X?"
115 INPUT X
120 V = X*(50 - 2*X)*(60 - 2*X)
125 PRINT "VOLUME IS"; V
130 GOTO 110
```

Before leaving this problem it is worth noting that since equation (3.3) is a cubic it could not be solved by methods usually taught on an elementary course. We should also say that the solution so far suggested leads to a rather shallow box. Since a cubic can have other solutions, there may be a deeper one that is possible, and we leave it to the reader to find out if an alternative, and more reasonably dimensioned, box is possible.

3.6 THE PROBLEM OF INFLATION

In Great Britain in December 1975 the retail price index was 146 and by December 1981 it had risen to 308.8 which is an increase of 111.5% in 6 years. Superficially this looks like 111.5/6% or 18.5% per year; no wonder so much fuss is made about the effects of inflation! However, we ought to look a bit more deeply into this situation. Inflation is usually measured by the percentage increase over one year, and in the above calculation we are looking for the average rate over six years. Suppose this rate is x% then in the first year the index will increase by

$146 \times (x/100)$ to $146 \times [1 + (x/100)]$, so it is increased by a factor of $1 + (x/100)$. For each of the six years the index is multiplied by the same factor so the rate $x\%$ that we are seeking satisfies the equation

$$146 \times \left(1 + \frac{x}{100}\right)^6 = 308.8 \ . \tag{3.4}$$

So far so good. The statement of the problem as an equation is within the capacity of a student of elementary algebra, but the solution cannot be found with techniques that are usually met on such a course. Many teachers would therefore avoid such a problem, however, with the aid of a computer, and the kind of method based on trial and error that is outlined above, the solution is accessible to elementary students. Progam 3.4 with lines 120 and 125 changed to

```
120  I = 146*(1 + X/100)↑6
125  PRINT "INDEX IS"; I
```

is all that is required. As before one can quickly find a value of x that is too small and another that is too large and then reduce the interval containing the root until a reasonable accuracy is obtained. It is interesting to compare the result with the value of 18.58% obtained by the initial calculation. It is instructive not only to comment on how far wrong this initial value was, but also to discuss why the answers should be different. This kind of problem is repeated in many situations where a rate of growth or decline is involved. Examples may be found in population change, radioactive decay, etc.

It is interesting to compare the average rate of inflation calculated by the above method with the published figures for the cost of living index for each of the six years, and the rate of inflation obtained from them. A straight arithmetical average of the rates in Table 3.2 may be compared with the result obtained as a solution to equation (3.4). This leads to a discussion of why the figures are so similar, and to the question of whether or not they will always be so.

Table 3.2– Retail price index for Great Britain.

Year	1975	1976	1977	1978	1979	1980	1981
Great Britain retail price index in December	146.0	168.0	188.4	204.2	239.4	275.6	308.8
% rise over the previous year (inflation rate)		15.1	12.1	8.4	17.2	15.1	12.0

3.7 REFINING THE METHOD OF SOLUTION

Both of the above problems led to equations that could not be solved in elementary algebra, and in both cases a solution could be found by first finding an interval containing a root and then reducing the size of the interval until the desired accuracy is obtained. A refinement of this approach leads to the interval bisection (or binary chop) method for solving equations numerically, and a description of it will be found in Chapter 13. It is normally best to restate the equation in the form $f(x) = 0$ then a standard program can be used to find the required solutions whatever the problem. If computer programming forms part of the course being followed then it is perhaps better still to start with some problems such as those given here and lead to a development of the method including the necessary programming.

3.8 THE FIELD PROBLEM

This is not so much a realistic problem as a source of mathematical exploration and the development of ideas. The problem is deliberately simple so that it is easily grasped and attention can be fully focused on methods of solution.

The problem is to find the measurements of a rectangular field using exactly 100 m of fence and having an area of 500 m². The unknowns are the length L and breadth B and the problem may be represented by the two equations

$$L + B = 50 \quad (3.5a)$$

$$LB = 500 \quad (3.5b)$$

These may of course be solved simultaneously, and the elimination of one unknown leads to a quadratic. There thus seems to be two solutions to the problem, but examination of these reveals just what one would expect of equations that are symmetrical in L and B. It is other approaches to the solution that is our concern here, and we believe that some, if not all, of these are accessible to those who might find the manipulation required for the elimination approach too damanding.

First consider a straight trial and error approach leading to an interval containing the solution. This may be conveniently tabulated as in Table 3.3.

Table 3.3.

L	B	Area $= L \times B$	Comment on area	Interval containing the value of L
30	20	600	Too large	
25	25	625	Larger still	
35	15	525	Too large but the best yet	
40	10	400	Too small	$35 < L < 40$
37	13	481	Too small	$35 < L < 37$
36	14	504	Too large	$36 < L < 37$
36.5	13.5	492.75	Too small	$36 < L < 36.5$

Values of L are chosen and B is obtained from equation (3.5a). The area is then evaluated and compared with the value required to fit equation (3.5b). It does not take very long to continue the table to the point where reasonable values have been obtained. The computer can easily be programmed to carry out the steps automatically, and an example is given as Program 3.5 (below). The first section starts with L1 equal to 25 and L2 is one more than L1. B1 and B2 are the corresponding values of B, A1 and A2 the areas, and E1 and E2 the amounts by which the areas are in error. In line 140 L1 in increased by 1 and the search continues until the test in line 135 detects errors of opposite sign thus indicating an answer between L1 and L2. Control now passes to line 200 and the interval is progressively halved in length by taking LM midway between L1 and L2. The test in 225 determines whether L1 or L2 takes the value of LM. The process continues until the test in 235 detects a small enough interval.

Program 3.5 – To solve the field problem

```
100 REM**THE FIELD PROBLEM**
110 REM**SEARCH FOR INITIAL INTERVAL**
115 L1 = 25
120 B1 = 50 - L1 : A1 = L1*B1 : E1 = 500 - A1
125 L2 = L1 + 1 : B2 = 50 - L2 : A2 = L2*B2 : E2 = 500 - A2
130 IF E1*E2 < 0 THEN GOTO 200
140 L1 = L1 + 1 : GOTO 120
200 REM**REDUCE INTERVAL**
210 LM = (L1 + L2)/2 : BM = 50 - LM
215 AM = LM*BM : EM = 500 - AM
220 IF E1*EM < 0 THEN L2 = LM : A2 = AM : GOTO 230
225 L1 = LM : A1 = AM : E1 = EM
230 IF ABS(L2 - L1) > 0.0001 THEN GOTO 210
240 PRINT "LENGTH OF FIELD IS BETWEEN"
241 PRINT L1;" AND "; L2
```

A second approach uses equations (3.5a) and (3.5b) alternately. Start with some chosen value for B, call this B_0, then equation (3.5a) gives a corresponding value L_0 for L. Use L_0 in (3.5b) to give a new value B_1 for B. This, in turn, is used in (3.5a) to give L_1. We continue in this way to obtain a sequence of values for B and L. Of course we have no assurance that the process will lead to a solution, but if it does eventually lead to the value B^* and a corresponding L^* such that B^* in (3.5a) gives L^* and L^* in (3.5b) gives B^* again, then this must be a solution. In practice the process would be stopped when the values fit to a sufficient accuracy. Suppose we take $B_0 = 25$ then the following values are obtained:

$B_0 = 25$ $\quad L_0 = 25$
$B_1 = 20$ $\quad L_1 = 30$
$B_2 = 16.66$ $\quad L_2 = 33.33$
$B_3 = 15$ $\quad L_3 = 35$
$B_4 = 14.29$ $\quad L_4 = 35.71$
$B_5 = 14$ $\quad L_5 = 36$
$B_6 = 13.89$ $\quad L_6 = 36.11$
$B_7 = 13.85$ $\quad L_7 = 36.15$

The values are now changing quite slowly and a few more steps leads to answers that are quite accurate enough. Progam 3.6 (below) can be used to perform the iteration automatically. In line 120 a copy is made of the value of B in BP so that in line 140 two successive values can be tested and the process repeated until insignificant further changes are taking place.

Program 3.6 – An iterative solution of the field problem

```
100 REM**FIELD PROBLEM ITERATION 1**
110 PRINT "WHAT STARTING B";
115 INPUT B
120 BP = B
125 L = 50 - B
130 B = 500/L
135 PRINT "B = "; B; "L = "; L
140 IF ABS(B - BP) > 0.0001 THEN GOTO 120
```

This is an example of an iterative process that converges, but convergence will not always occur for such processes. This leads to questions about the conditions that are necessary for the process to work. Though a full discussion of the conditions for convergence may well have to be postponed, the experience gained is valuable as preparatory work. Variations that may be investigated include the use of different values for B_0. Does the process work for any starting value? Does it make any difference if the equations are used in reverse order, that is, use (3.5b) to obtain L_0 from B_0 then (3.5a) to get B_1 from L_0, etc.?

In a third approach the problem is restated by using (3.5a) to write $B = 50 - L$ then substitute for B in (3.5b) to give

$$L(50 - L) = 500$$

This then gives

$$L^2 = 50L - 500 \tag{3.6}$$

and we see at once that $L > 10$. Since we know that L is positive we may write

$$L = \sqrt{50L - 500} \tag{3.7}$$

Equation (3.7) is the sort of erroneous attempt at an analytic solution that confused students sometimes make. Here it can form the basis of another attempt at an iterative solution. If the correct solution is substituted into the right-hand side of (3.7) then the value obtained is the solution itself. If any other value is substituted then a different number will be generated. We may start by choosing a value L_0 then $L_1 = \sqrt{50L_0 - 500}$. Now repeat with L_1 so that $L_2 = \sqrt{50L_1 - 500}$. The whole process is represented by the recurrence relation

$$L_{n+1} = \sqrt{50L_n - 500} \tag{3.8}$$

Starting with the choice $L_0 = 20$ the first few values obtained are:

$$L_0 = 20$$
$$L_1 = 22.36$$
$$L_2 = 24.86$$
$$L_3 = 27.26$$

This does not look as though it is converging to a solution but the sequence ought to be taken much further before we can be sure. A simple computer program such as Program 3.7 (below) can help here. The program should be tried with a variety of starting values to see how it performs. Line 140 is a delay to prevent the values being printed too quickly and program execution can be halted with the stop key.

Of course the rearrangement of the problem to give recurrence relation (3.7) is not the only one that is possible. Equations (3.5) can be rearranged to give $L = 50 - B$ and $B = 500/L$. Substituting for B in the first of these gives

$$L = 50 - \frac{500}{L}$$

and this is the basis for the recurrence relation

$$L_{n+1} = 50 - \frac{500}{L_n} \tag{3.9}$$

This may now be tried in the same way and once again we note that $L > 10$ since L must be positive. Program 3.7 needs only line 135 to be changed to the formula given by equation (3.9). Again we can investigate various starting values. It is recommended that both this and the pevious iterative method be tried. The differing results raise a number of questions about why the methods behave in the way that they do. In both cases one dimension of the required rectangle is obtained and the other must then be found from equation (3.5a).

Program 3.7 – An attempt at an iterative solution of the field problem

```
100 REM** FIELD PROBLEM ITERATION 2**
110 PRINT "WHAT STARTING VALUE";
115 INPUT L
120 N = 0
125 PRINT N, L
130 N = N + 1
135 L = SQR(50*L - 50)
140 FOR K = 1 TO 1000: NEXT K
150 GOTO 125
```

Yet another iterative method results if equation (3.6) is rearranged to give

$$25L = L^2 - 25L + 500$$

or

$$L = \frac{L^2 - 25L + 500}{25}$$

and this leads to the recurrence relation

$$L_{n+1} = \frac{L_n^2 - 25L_n + 500}{25} \tag{3.10}$$

This can be applied in the same way as before. Not all starting values will work, and it may be worth while finding out experimentally the range of convergence.

No doubt the reader will be able to suggest other re-arrangements of the formulae leading to yet more attempts at an iterative solution. Other equations may be tackled in a similar manner. In the case of the example used here an analytical solution is possible and may be compared with the results obtained iteratively. In other cases an analytical solution may not be possible and it is in those cases that iterative methods come into their own. By the use of such methods the range of problems that may be tackled in elementary courses is greatly extended. We believe also that a wider experience of algebra as a language can be gained, without lengthy pieces of manipulation that so often detract from an interest in the subject and lead to frustrating slips in working.

More efficient iterative methods are of course available, and some of these are given in Chapter 13, but often a more extensive mathematical basis than is assumed in this chapter is required to understand why they work.

In general, then, with the aid of a computer we can easily explore functions and equations that would be too complex for manipulative and analytical techniques to be applied. We feel, strongly, that a mathematics syllabus in which pupils only meet the very restricted class of functions and equations that can be tackled within the techniques of elementary manipulative algebra, does a grave disservice to the claims of mathematics to be widely applicable to the 'real world'. For the student who intends to continue with mathematical studies, the experience of a range of complicated functions and functions of several variables provides a much firmer base for understanding the generalisations of analysis.

4

Graphing by computer

Microcomputers which produce their output on a TV screen or monitor can be programmed to display a particular character at a given position on the screen. This allows us to produce graphs of functions, of more or less sophistication, on virtually all modern microcomputers. Unfortunately, though, the way in which microcomputers organise their display varies widely between machines as do the commands that they use. Fortunately, though, the principles underlying graph plotting are quite straightforward and can be easily adapted to produce acceptable results on most machines. We shall describe three kinds of graphics systems of increasing sophistication – text characters, teletext graphics and pixel based (high resolution) graphics. In each case a character can be displayed at a rectangular area of the screen specified by a pair of co-ordinates giving its screen column and row relative to an origin that is either at the bottom left-hand corner or top left-hand corner of the screen.

4.1 TEXT CHARACTERS

Each printable character of a microcomputer has a code number between 0 and 255. Many microcomputers use the ASCII system (American Standard Code for Information Interchange) and even those that do not follow a very similar system. Thus the usual code for the letter 'A' is 65, 'B' is 66, 'C' is 67 and so on. A useful symbol for plotting graphs is the multiplication sign '*' which usually has code 42 (so perhaps that was the question in *The Hitch Hiker's Guide to the Galaxy*!).

There are also (see Table 4.1) codes for the special non-printing characters which perform operations such as (i) clearing the screen (usually code 12); (ii) starting a new line – equivalent to the RETURN key (usually 13); (iii) moving the cursor up (11), down (10), right (9), left (8) and to the top left-hand corner – sometimes called HOME (30); and (iv) to delete a character – once called 'backspace and rubout' (127).

Table 4.1

Micro	Clear	Return	Up	Down	Right	Left	Home	Delete
BBC	12	13	11	10	9	8	30	127
380Z	12/31	13	11	10	18	8	29	127
PET	147	13	145	17	29	157	19	20

The screen is usually about 40 characters wide and 25 lines deep – though some of the more recent, cheaper, micros use smaller values and some of the 'business-orientated' machines have a width of 80 characters. To cope with this we shall introduce the variables SW for 'screen width' and SH for 'screen height' and assume that the 'origin' (0, 0) corresponds to the top left-hand corner.

In order to plot a character at a given position on the screen there are, again, several different approaches of which we shall illustrate just three. Again we have to hope that the reader can adapt at least one of these to his own machine. The approaches are (i) use of a single command such as PLOT X, Y, Z (380Z) or PRINT TAB(X, Y); CHR$(Z) (BBC) or PRINT AT X, Y; CHR$ Z (Spectrum), (ii) use of cursor control codes to position the cursor at the desired place on the screen and (iii) putting the code for the character directly into the memory (RAM) location used for the appropriate position on the screen – known as 'poking to the screen'. In each case we shall develop a subroutine to perform the task of plotting the character whose code is C at the screen position in the Xth column and Yth row. It will be assumed that values for SW and SH (screen width and height) have already been defined.

4.1.1 Direct commands

On the BBC micro there are screen modes, such as MODE 6 and MODE 7 in which the screen is 40 characters wide and 25 lines deep – other modes use 20, 40 or 80 characters wide and 25 or 32 characters deep. The 'two-dimensional' cursor tabulation function TAB(X, Y) takes the cursor to the area of the screen in the Xth column and Yth row when used as part of a PRINT statement

Program 4.1 – An example of direct text plotting

```
100 REM PROGRAM 4.1 – EXAMPLE OF DIRECT TEXT
    PLOTTING ON BBC
110 MODE 7
120 SW = 39 : SH = 24
130 INPUT X,Y,C
```

```
 140  GOSUB 1000
 150  END
1000  REM PLOT C AT (X, Y) DIRECTLY
1010  IF X<0 OR X>SW OR Y<0 OR Y>SH THEN RETURN
1020  PRINT TAB(X, Y); CHR$(C);
1030  RETURN
```

Line 1020 can also be entered as: VDU 31, X, Y : VDU C or even VDU 31, X, Y, C and other computers such as the Research Machines 380Z use PLOT X, Y, C.

4.1.2 Cursor control

On the Commodore PET, for example, there are control codes to move the cursor. If we introduce string variables H$, D$, and R$ to stand for cursor HOME, DOWN and RIGHT then we can arrive at the required position (X,Y) by printing H followed by X repetitions of R$ and Y repeats of D$. To make the programming easier it is preferable to set D$ to be a string of SH repetitions of 'cursor down' and R$ to be a string of SW repeats of 'cursor right' so that we can produce the desired effect by:

```
PRINT H$; LEFT$(R$,X); LEFT$(D$,Y); CHR$(C)
```

using the LEFT$ function to take the substring of the required length.

Program 4.2 – An example of the use of cursor control

```
 100  REM PROGRAM 4.2 – EXAMPLE OF CURSOR CONTROL
      ON PET
 110  PRINT CHR$(147) : REM CLEAR SCREEN
 120  H$ = CHR$(19) : REM HOME CURSOR
 130  SW = 39 : SH = 24 : REM SCREEN WIDTH & HEIGHT
 140  D$ = "" : R$ = "" : REM THE 'NULL STRING'
 150  FOR I = 1 TO SH : D$ = D$ + CHR$(17) : NEXT I
 160  FOR I = 1 TO SW : R$ = R$ + CHR$(29) : NEXT I
 170  INPUT X,Y,C
 180  GOSUB 1000
 190  END
1000  REM PLOT C AT (X,Y) – CURSOR CONTROL
1010  IF X<0 OR X>SW OR Y<0 OR Y>SH THEN RETURN
1020  PRINT H$; LEFT$(R$,X); LEFT$(D$,Y); CHR$(C);
1030  RETURN
```

4.1.3 Poking to the screen

Many computers have what are known as 'memory-mapped screens'. This means that there is a portion of the memory (RAM) whose contents are in direct correspondence to the characters displayed and hence altering the contents of one of these locations will cause the display to change. The way in which such storage is organised varies between machines. First it is necessary to establish the mapping between the position (X, Y) on the screen and the address AD of the location in RAM which corresponds to it. For the PET and BBC micros this is very straightforward and only marginally more complex for the 380Z – however, the Apple is rather more troublesome. The most direct approach to altering memory is to use the POKE command. POKE A, B causes the memory location A (where, usually, $0 \leqslant A < 65536$) to be altered to the 8-bit binary pattern (byte) equivalent to the decimal number B where $0 \leqslant B < 256$. The procedure then involves first finding the decimal number of the location corresponding to the top left-hand corner of the screen – the so-called *base address* BA (which is 32768 on the PET and, in MODE 7 for instance, is 31744 on the BBC micro). Next the address of the memory location of the desired screen position is found. For the PET and BBC micros in 40 column mode this consists of adding 40Y + X to the base address. Finally the code for the text character is inserted in this location by a POKE statement.

Program 4.3 – An example of poking to the screen

```
 100  REM PROGRAM 4.3 – EXAMPLE OF POKING TO THE
      SCREEN ON PET
 110  PRINT CHR$(147) : REM CLEAR SCREEN
 120  SW = 39 : SH = 24 : REM SCREEN WIDTH AND HEIGHT
 130  BA = 32768 : REM BASE ADDRESS
 140  INPUT X,Y,C
 150  GOSUB 1000
 160  END
1000  REM PLOT C AT (X,Y) BY POKING
1010  IF X<0 OR X>SW OR Y<0 OR Y>SH THEN RETURN
1020  AD = BA + X + 40*Y
1030  POKE AD,C
1040  RETURN
```

4.2 TELETEXT CHARACTERS

With the advent of visual information systems such as PRESTEL, TELETEXT and VIEWDATA which include, for example, the BBC CEEFAX and IBA ORACLE broadcasts, a standard for text characters, graphic symbols and colour range has been introduced in much the same way as the telegraph, ticker-tape and Telex gave rise to the ASCII standard for information interchange. The screen is considered to consist of a rectagular grid of positions each of which can hold a symbol which might be one of the familiar keyboard character set or might be a graphical symbol composed of 6 blocks arranged in a 2 by 3 pattern:

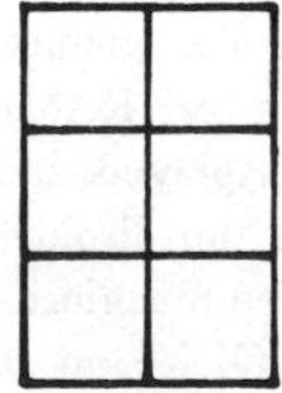

The usual convention is that within any one such printing position there is only one foreground colour and one background colour. As each of the small blocks can be in one of the two different colours there are $2^6 = 64$ different combinations possible. Each of these possible combinations will be given a code number for transmission. This is usually constructed by allocating a power of 2 to each small block, adding up the powers corresponding to the blocks to be displayed in the current foreground colour and adding on some fixed quantity (called an 'offset') to put the code into the desired part of the range 0–255 of allowable codes.

For example in the BBC microcomputer's Teletext mode – MODE 7 – the offset is 160 and the powers of 2 are arranged as shown:

1	2
4	8
16	64

Thus the symbol:

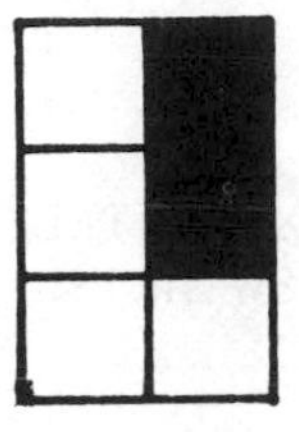

– an L-shaped arrangement of 4 foreground blocks (shown in white) and 2 background blocks (shown in black) has the code $1 + 4 + 16 + 64 + 160 = 245$ and (in the right circumstances) the instruction PRINT CHR$(245) will display the symbol. Most microcomputer implementations of Teletext graphics need some extra code or command to access the text and graphics character set rather than the conventional 'text only' set. In the BBC machine a control code to select the graphics foreground colour has to be given first – these are in the range 145–151 and a white L-shaped symbol will be produced by:

```
MODE 7 : PRINT CHR$(151); CHR$(245) .
```

Continuing with the example of the BBC's MODE 7 the text screen has 40×25 printing positions but the use of Teletext graphics gives us 80×75 small block available. In order to plot a small block we need a three-stage process. First to find which text position the small block lies within and also find the appropriate power of 2 for the position of the block within the array. Compute the code for the Teletext graphics symbol consisting of just this one small block in foreground colour with the other five blocks in background colour. Next find the code of the graphics symbol already displayed at the appropriate position on the screen – this can be done by a PEEK to the corresponding location in RAM or by using a special routine supplied by the manufacturer which is known as an 'Operating System call'. Finally the two codes are combined with the logical OR operation to give the code of the combined Teletext graphics character which is then written back into the RAM location by a POKE or an Operator System call or placed on the screen by a PRINT statement. A description of how to do this on the BBC micro appears in the manual (p. 158) and so will not be repeated here.

With the Research Machines version of BASIC this procedure has been much simplified. The command GRAPH will clear the top 20 lines of the screen ready for both text and Teletext graphics symbols. The command PLOT X,Y,2 will plot a small white Teletext block in any position in this area where $0 \leqslant X \leqslant 79$ and $0 \leqslant Y \leqslant 59$ with the point (0,0) at the top left-hand corner. Thus the 'system' does all the hard work of computing code, peeking, ORing, and poking – which makes life a lot easier.

Program 4.4 – An example of plotting with Teletext characters

```
 100 REM PROGRAM 4.4 – EXAMPLE OF 380Z TELETEXT
     PLOTTING
 110 GRAPH : REM SET GRAPHICS AND CLEAR SCREEN
 120 SW = 79 : SH = 59 : REM SCREEN WIDTH AND HEIGHT
 130 INPUT X,Y
 140 GOSUB 1000
 150 END
1000 REM PLOT AT (X,Y) – TELETEXT
1010 IF X<0 OR X>SW OR Y<0 OR Y>SH THEN RETURN
1020 PLOT X,Y,2
1030 RETURN
```

Research Machines also include the command LINE X,Y,2 to plot a rough line of Teletext small blocks from the last point plotted to the point (X,Y). If PLOT X,Y,1 or LINE X,Y,1 is used then the small blocks will be shown in a foreground colour of grey rather than white. In GRAPH mode the bottom four lines of the screen are allowed to scroll like the street names underneath a town map. When a program using GRAPH finishes it is desirable to cancel this mode by using the TEXT command, otherwise unreadable things happen when an attempt is made to list the program!

4.3 HIGH-RESOLUTION PLOTTING

Here the screen is divided up into a rectangular array of small rectangles, called *pixels*. Any pixel (X,Y) can be plotted by commands such as:

```
HPLOT X,Y          – Apple
PLOT 69,X,Y        – BBC
CALL "PLOT",X,Y    – RML
```

The smaller the size of the pixel the higher the resolution. Typical values for the numbers of pixels on the screen are:

280 × 160	– Apple
640 × 256, 320 × 256, 160 × 256	– BBC
320 × 192, 160 × 96	– 380Z

4.4 A SINE CURVE – SOME EXAMPLES OF PLOTTING

As an example of curve plotting consider producing the graph of one wave of the sine function $y = \sin x$ for $-\pi \leqslant x \leqslant \pi$. Here the x range is 2π and the y range is 2. We want the centre of the screen block in the top left-hand corner of the screen to correspond to $(-\pi, 1)$ and the centre of the block at the bottom right to be equivalent to $(\pi, -1)$. This means that if SW and SH are set to the screen width and height and the screen origin is the *top* left-hand corner then we seek a mapping that gives:

$$
\begin{aligned}
(-\pi, 1) &\rightarrow (0,0) \\
(\pi, -1) &\rightarrow (\mathrm{SW}, \mathrm{SH}) \\
(x, y) &\rightarrow (\mathrm{X}, \mathrm{Y})
\end{aligned}
$$

which yields:

$$x = \frac{2\pi}{\mathrm{SW}+1}(\mathrm{X} - \mathrm{SW}/2)$$

$$\mathrm{Y} = (\mathrm{SH}/2)(1 - y)$$

If the screen origin is the bottom left-hand corner then a few of the signs will need to be changed and we arrive at $\mathrm{Y} = (\mathrm{SH}/2)(1 + y)$. In each case, if SW or SH is odd then we take the nearest integer to SW/2 or SH/2.

The procedure, then, is to step X from 0 to SW one block at a time, to transform X to x, to calculate y corresponding to x and, finally to transform y to Y. We give a number of programs to illustrate how the procedure may be implemented on several microcomputers in a variety of graphics styles to show how easy and versatile such a program is.

Program 4.5 – A sine curve using text characters

```
100 REM PROGRAM 4.5 – TEXT PLOTTING ON BBC
110 MODE 7
120 FOR X% = 0 TO 39
130    X = (X% - 19)*PI/20
140    Y = SIN(X)
150    Y% = 12*(1 – Y)
160    PRINT TAB(X%,Y%); CHR$(170);
170 NEXT X%
```

Here MODE 7 is used which takes up very little memory (1K) and has a 40 X 25 character display. The curve is plotted using the asterisk symbol * which has the ASCII code 170. As the screen positions are integers the program illustrates the use of integer variables X% and Y%.

Program 4.6 – A sine curve by poking to the screen

```
100  REM PROGRAM 4.6 – TEXT PLOTTING ON PET
110  PRINT CHR$(147);
120  BA = 32768
130  FOR SX = 0 TO 39
140     X = (SX - 19)*π/20
150     Y = SIN(X)
160     Y% = 12*(1 - Y)
170     POKE(BA + SX + 40*Y%),42
180  NEXT SX
```

Here the screen code for the asterisk, 42, is poked directly into the screen memory which starts at location 32768. 147 is the code for the button which clears the screen. Although the PET does possess integer variables they may not be used in FOR – NEXT loops to count the loop.

Program 4.7 – A sine curve using Teletext characters

```
100  REM PROGRAM 4.7 – TELETEXT PLOTTING ON
     RML 380Z
110  GRAPH
120  FOR SX = 0 TO 79
130     X = (SX - 39)*3.14159/40
140     Y = SIN(X)
150     SY = 29*(1 - Y)
160     PLOT SX, SY, 2
170  NEXT SX
180  TEXT
```

As the RML BASIC does not allow integer variables the screen co-ordinates are held in SX and SY.

Program 4.8 – A sine curve using high resolution graphics on BBC

```
100  REM PROGRAM 4.8 – HIGH RESOLUTION PLOTTING
     ON BBC
110  MODE 4
120  FOR X% = 0 TO 1279 STEP 4
130     X = (X% - 639)*PI/640
140     Y = SIN(X)
150     Y% = 511*(1 + Y)
160     PLOT 69, X%, Y%
170  NEXT X%
```

As the BBC graphics modes use a notional graphics grid of 1280 X 1024 points but the screen itself is composed in MODE 1 of 320 X 256 pixels there is not a one-to-one mapping between points X%,Y% and pixels. Hence (0,0), (3,0), (0,3) and (3,3) all refer to the same pixel in the bottom left-hand corner of the screen. To make the procedure more efficient, then, we need only step the X% variable by 4 to cover the full range of horizontal pixels in this case.

Program 4.9 – A sine curve using the resolution graphics on 380Z

```
100 REM PROGRAM 4.9 – HIGH RESOLUTION PLOTTING
    ON 380Z
110 CALL "RESOLUTION",0,2
120 FOR SX = 0 TO 319
130    X = (SX - 159)*3.14159/160
140    Y = SIN(X)
150    SY = 95*(1 + Y)
160    CALL "PLOT", SX,SY,3
170 NEXT SX
```

For the Research Machines 380Z to display high-resolution graphics an extension board has to be fitted. There are high-resolution (320 X 192) and medium-resolution (160 X 96) modes and the display can plot points in a variety of colours or shades. The 3 digit in the command CALL "PLOT",X,Y,3 denotes that the point is to be plotted at position (X,Y) in the colour whose code is 3 – in this case white.

Program 4.10 – A sine curve using the resolution graphics in Apple

```
100 REM PROGRAM 4.10 – HIGH RESOLUTION PLOTTING
    ON APPLE
110 HGR
120 HCOLOR = 3
130 FOR SX = 0 TO 279
140    X = (SX - 139)*3.14159/140
150    Y = SIN(X)
160    Y% = 79*(1 - Y)
170    HPLOT SX,Y%
180 NEXT SX
```

4.5 GENERAL TECHNIQUES FOR PLOTTING

A more general program is one that will display the graph of a given function for a given range of values of the independent and dependent variables x,y. Ideally we want a program that will allow a pupil to enter his or her own function and ranges and for the curve to be drawn as a series of line segments joining adjacent points on the curve. We do not want the program to be halted (or to 'crash') because the function is not defined at some part of the range, for example Y = SQR(X) will cause an error message to be printed if X = –1 as does Y = (X + 3)/(X – 3) at X = 3. Also some computers produce error messages if an attempt is made to plot a point that is off the screen. The program will be developed using high-resolution graphics in BBC BASIC but many of the techniques are applicable to other computers.

Program 4.11 – A general graph plotter

```
 100 REM PROGRAM 4.11 – A GENERAL GRAPH ROUTINE
     FOR BBC
 110 DEF FNA(X) = SQR(X)/(X - 2)
 120 INPUT "XLOW, XHIGH", XL, XH
 130 INPUT "YLOW, YHIGH", YL, YH
 140 NP = 40 : XR = XH - XL : YR = YH - YL
 150 IF XR <= 0 OR YR <= 0 THEN GOTO 120
 160 XS = XR/NP : SH = 1023 : SW = 1279
 170 MODE 4 : PF% = -1 : XM = XL
 180 ON ERROR GOTO 1000
 190 FOR X = XM TO XH STEP XS
 200   X% = SW*(X - XL)/XR
 210   Y = FNA(X)
 220   Y% = SH*(Y - YL)/YR
 230   IF X% < 0 OR X% > SW OR Y% < 0 OR Y% > SH THEN
       PF% = -1 : GOTO 260
 240   IF PF% THEN PLOT 69, X%, Y% ELSE DRAW X%, Y%
 250   PF% = 0
 260 NEXT X
 270 END
1000 REM ERROR HANDLING
1010 IF ERR = 17 THEN PRINT "Interrupted" : END
1020 IF ERR = 23 OR ERR = 18 OR ERR = 24 OR ERR = 22 OR
     ERR = 21 OR ERR = 20 THEN GOTO 1040
1030 REPORT : END
1040 PF% = -1 : XM = X + XS : GOTO 190
```

This program could benefit from being made both more efficient and more elegant but it serves to illustrate techniques that can be applied to several other common microcomputers that do not have the structural advantages of BBC BASIC. The function is defined at line 110 and this can be altered to plot any function you like, XL and XH hold the lower and upper limits for the independent variable x and YL, YH do the same for the dependent variable y. XR and YR hold the range for each variable and NP is the number of points to be plotted. If silly ranges for x or y are entered then they are 'trapped' at line 150 and handled in a rather crude way. XS holds the step by which x will be incremented and SH, SW hold the screen height and width. In this version of graphics the origin of screen co-ordinates is at the lower left-hand corner. PF% is an integer variable which is used as a 'flag' to determine whether a new point should be plotted (signalled by PF% = –1: the computer's code for 'true') or if a line should be drawn from the last point visited to the new point (shown by PF% = 0: the code for 'false').

Line 180 is a 'trap' to stop the computer from interrupting the program and giving its own error messages when 'run-time' errors are encountered. From this point in the program onwards if an error is encountered the program will jump to our own error-handling routine at line 1000. If the rest of the program is correct then the only errors to which we want to give special attention will arise at line 210 which refers to the definition of FNA(X) in line 110. The main loop takes each value of X and maps it onto a screen value X%, calculates the corresponding value of Y (where possible) and maps that onto a screen value Y%. The corresponding screen point (X%, Y%) is then checked at line 230 to see if it is on the screen. If it is not then the plotting flag PF% is set to 'true' so that the next time a point on the screen is calculated it will be plotted rather than have a possibility illegal or incorrect line drawn to it.

Line 240 plots the point or draws a line to it depending upon the state of the flag PF%, and, because it is a 'legal' point the plotting flag is cleared to false in readiness to draw a line to the next point computed if it, too, is found to be on the screen.

The error-handling routine at line 1000 uses the computer's own variable ERR which is set when an error is detected. BBC BASIC has a variable ERL that is set to the line number of the command which caused an error and ERR is set to the code which specifies the kind of error. We are particularly concerned with three types of error message. If the person at the keyboard presses the ESCAPE key to interrupt the program then this action is usually treated as an error which has code 17 – thus line 1010 checks for an interruption. The next class of errors are all computational – we want to see if the function FNA(X) was not properly defined for the current value of X. The computational errors that are checked for in line 1020 are: (i) 'Accuracy lost' (code 23) – usually encountered in working out trigonometric functions of very large angles; (ii) 'Exp range' (24) – met in evaluating EXP(X) if X > 88; (iii) 'Division by zero' (18) – which is self-explanatory; (iv) 'Log range' (22) – met in evaluating the logarithm of a negative number; (v) 'Negative root' (21) – if the square root of a negative number is attempted; and (vi) 'Too big' (20) – if a calculation results in a number greater than the maximum the computer can hold. In each of these cases we jump to line 1040 to set the plotting flag PF% to 'true' showing that the next 'legal' point is to be plotted. We now wish to return to the main part of the program

Unfortunately the process of error-recognition in this version of BASIC causes the program to 'lose its place' in FOR–NEXT loops, REPEAT–UNTIL loops, subroutines, procedures and functions. Thus we cannot jump back to the end of the current X loop at line 260. However, the values of all variables are preserved so all we have to do is to set the value of XM to the next value of X that would be encountered, that is, X + XS and re-enter the program at the start of the loop at line 190.

If the error was neither an interruption nor an arithmetic error than it must be the result of a bad definition of the function FNA(X) in line 110 and so we need to report the error and halt the program to allow it to be rectified.

A powerful feature of this version of BASIC not generally implemented yet in other micros is the ability to enter a function from the keyboard while the program is running and to evaluate it to get a numeric value using the EVAL function. Try the following amendments to see the effect of this.

```
110  INPUT "Please type of your function of X : Y = " F$
210      Y = EVAL(F$)
```

Obviously there are many possible extensions to this program, such as plotting several functions in different colours, automatic calculation of appropriate values of YL and YH corresponding to given values of XL and XH, marking of axes, labelling of scales, options to replot the same function over different X and Y ranges and so on. Undoubtedly a graph-plotting program is a most versatile tool for the mathematics teacher. Thc program we have illustrated plots Y as an explicit function of X but very few modifications are needed for it to plot a curve from a parametric definition: $x = f(t)$, $y = g(t)$ for a given range of the parameter t. Consider, for example, plotting the astroid $x = \cos^3 t$, $y = \sin^3 t$ for $0 \leqslant t \leqslant 2\pi$. We could define FNX(T) and FNY(T) each on separate line just as FNA(X) was defined at line 110. BBC BASIC also allows multi-line function definitions to allow more complex functions to be defined so we will use this example to illustrate the technique. Such function definitions have to come at the end of the program. The required amendments are thus:

```
 110
 135 INPUT "TLOW, THIGH", TL, TH
 155 TR = TH - TL : IF TR <= 0 THEN GOTO 135
 160 TS = TR/NP : SH = 1023 : SW = 1279
 190 FOR T = TL TO TH STEP TS
 195    X = FNX(T)
 210    Y = FNY(T)
 260 NEXT T
1050 PF% = -1 : TL = T + TS : GOTO 190
2000 DEF FNX(T)
2010 C = COS(T)
2020 = C*C*C
2030 DEF FNY(T)
2040 S = SIN(T)
2050 = S*S*S
```

The extension to plotting polar forms $r = f(\theta)$ is also simple and is left to the reader to explore. Another interesting extension to try is the graphing of linear inequalities such as $3x + 4y < 60$.

5

Matrices and vectors

Both matrices and vectors are met nowadays in most elementary mathematics syllabuses. Many real applications of mathematics use matrices and it is commonplace now to use a computer for the necessary calculations. In most instances the amount of calculation to be done makes the use of a computer essential if one is not to be restricted to very simple cases. In the study of elementary mathematics, however, we are concerned to establish principles and develop understanding, and for this a computer is often not essential and in some cases not desirable. For example, the multiplication of two matrices needs to be experienced by hand in the introductory stages, but the number, size and complexity of problems that may be tackled can be increased if a computer is used. It is the purpose of this chapter to show some of the ways which a microcomputer may be used for matrix and vector work. It is our belief that it should not be used indiscriminately but that there are places in the current curriculum where its use can greatly assist the development of understanding.

First we will turn our attention to the input, output and storage of a matrix or vector. Unless they are very small, arrays will need to be used. We then consider the multiplication of two matrices or of a vector by a matrix. A numerical method for finding the inverse of a matrix completes the chapter. Matrices are also used in Chapter 6 on computer geometry.

5.1 INPUT, OUTPUT AND STORAGE OF A MATRIX

Matrices will be stored in arrays of two dimensions, the first being the number of rows and the second columns. Before a matrix can be used the computer needs to be told how big it is to be. This is done by a dimension statement. For example, suppose we wish to put the matrix

$$\begin{bmatrix} 5 & 4 & 3 & 2 & 1 \\ 6 & 5 & 4 & 3 & 2 \\ 7 & 6 & 5 & 4 & 3 \\ 8 & 8 & 8 & 4 & 4 \end{bmatrix}$$

into an array called A. Since there are 4 rows and 5 columns the dimension statement is

```
120  DIM A(4,5)
```

The computer now needs to be given the elements in a set order. It is quite possible to type the elements as the program is running by using an INPUT statement but we usually prefer to have the numbers in DATA lines. The advantage is that if a typing error is made then just one data line has to be changed and can be checked before the program is run. With keyboard input a mistake requires the program to be re-run from the start and for a large matrix the probability of getting it completely right may be quite low. Example Program 5.1 uses nested loops for the input. The rows are entered in turn and variable J keeps track of the row being read. K is then the column of the current element. Thus for each value of J variable K counts through the columns as the elements of row J are read. The elements of the matrix are read from the DATA lines 1010 onwards and for convenience one row of the matrix is in each data line.

Once the matrix is in the computer any element may be referred to by its row and column numbers. Thus in the example matrix A above, the value of A(3, 1) is 7 and of A(2, 4) is 3.

Outputting a matrix follows a similar pattern but there can be problems with reading the result if the row length is too large for the space available on one line of the output device. The VDU screen line is often split into four fields for output and one number is printed in each field. Thus the statement PRINT 5,6,7,8,9 results in the screen output:

```
5  6  7  8
9
```

Our example matrix would be displayed as:

```
5  4  3  2
1
6  5  4  3
2
7  6  5  4
3
8  8  8  4
4
```

which is a bit awkward to read. With less simple numbers it becomes more difficult, but by leaving a blank line between each line of the matrix and the next, readability is improved. On some microcomputers it is possible to vary the field size and this can be useful solution to the problem. Printers often use a default field size of ten characters so that, with 80 characters to a line, matrices with up to eight columns may be printed easily. Setting a different format is also a way of solving the problem.†

Program 5.1 – To input a matrix

```
 100
 100 REM**INPUT MATRIX**
 120 DIM A(4,5)
 130 FOR J = 1 TO 4
 135    FOR K = 1 TO 5
 140       READ A(J,K)
 145    NEXT K
 150 NEXT J
 999 END
1000 REM**DATA SECTION**
1010 DATA 5,4,3,2,1
1020 DATA 6,5,4,3,2
1030 DATA 7,6,5,4,3
1040 DATA 8,8,8,4,4
```

Program 5.2 – Lines to be added to Program 5.1 to output the matrix

```
900 REM**OUTPUT MATRIX**
910 FOR J = 1 TO 4
915    FOR K = 1 TO 5
920       PRINT A(J,K);
925    NEXT K
930    PRINT : PRINT
935 NEXT J
```

Program 5.2 outputs the example matrix to the VDU screen. In line 930 the first PRINT causes the output to move to the next line and the second PRINT gives a blank line to improve readability.

---

† On some microcomputers the printed number may exceed the default field size if sufficient figures are available. For example, 5.67, 82.96 causes no difficulty but 0.666666667, 0.333333333 have nine figures and occupy eleven places so that they exceed the field size. The result is that the 0.333333333 is printed in the third field instead of the second. A clearer output is obtained if numbers are truncated to say four decimal places for printing. In Program 5.2 line 920 can be modified to:

```
920 PRINT INT(A(J,K)*1000)/1000 ;
```

5.2 TO MULTIPLY A VECTOR BY A MATRIX

In studying transformations represented by matrices one sometimes needs to multiply a vector several times by the matrix or to multiply different vectors by the matrix. The process involves using the rows of the matrix in turn, each element of the row being multiplied by the corresponding element of the vector and the results added.

$$\text{Row 3} \rightarrow \begin{bmatrix} \\ a_{31}\; a_{32}\; a_{33}\; a_{34}\; a_{35} \end{bmatrix} \begin{bmatrix} v_1 \\ v_2 \\ v_3 \\ v_4 \\ v_5 \end{bmatrix} = \begin{bmatrix} x_1 \\ x_2 \\ x_3 \\ x_4 \end{bmatrix}$$

In the example

$$x_3 = a_{31} v_1 + a_{32} v_2 + a_{33} v_3 + a_{34} v_4 + a_{35} v_5 \;.$$

The sum on the right may be accumulated by using a loop with a counter K going from 1 to 5

```
455  X(3) = 0
460  FOR K = 1 TO 5
465     X(3) = X(3) + A(3,K)*V(K)
470  NEXT K
```

To compute all the components of vector x these program lines can be enclosed in another loop in which J counts from 1 to 4, the 3's of lines 415 and 425 being replaced by J. Often we shall be using sqaure matrices and example Program 5.3 includes a 3 × 3 matrix in the data section. The starting vector is input from the keyboard in lines 220 to 235. Line 225 is not essential but it is helpful to tell the user which component is required.

The matrix in the example is one in which the sum of each of the columns is one. Such matrices arise in situations where there is a fixed total that is of three kinds, but the proportions of each kind change with time in a predetermined way. For example the total might be the sales of a certain product, let us say shoe polish, that is made by three manufacturers A, B and C. The components of the vector give the sales of the three companies. The matrix contains information about which brand customers will buy next month according to which brand they brought this month. Column 1 means that of those who bought A this month 0.8 will again buy A next month, 0.1 will buy B and 0.1 will buy C. Similarly for those buying B now the proportions next month will be 0.3 for A, 0.4 for B and 0.3 for C. The last column applies in the same way to the purchasers of C. If the total number of customers is constant then each column of the matrix has a total of one.

Program 5.3 – To multiply a vector by a matrix

```
100  REM**INPUT MATRIX**
110  READ N
120  DIM A(N,N), V(N), X(N)
130  FOR J = 1 TO N
135    FOR K = 1 TO N
140      READ A(J,K)
145    NEXT K
150  NEXT J
200  REM**INPUT VECTOR**
210  PRINT "INPUT STARTING VECTOR"
220  FOR J = 1 TO N
225    PRINT "V(";J;") = ";
230    INPUT V(J)
235  NEXT J
300  REM**OUTPUT VECTOR**
310  PRINT
315  FOR J = 1 TO N
320    PRINT V(J);
325  NEXT J
330  PRINT
400  REM**MULTIPLY VECTOR BY MATRIX**
410  PRINT "DO YOU WANT TO MULTIPLY VECTOR BY
     MATRIX"
415  PRINT "YES OR NO";
420  INPUT A$
425  IF A$ = "NO" THEN END
430  IF A$ <> "YES" THEN GOTO 410
450  FOR J = 1 TO N
455    X(J) = 0
460    FOR K = 1 TO N
465      X(J) = X(J) + A(J,K)*V(K)
470    NEXT K
475  NEXT J
480  FOR J = 1 TO N
485    V(J) = X(J)
490  NEXT J
495  GOTO 300
1000 REM**DATA SECTION**
1005 DATA 3
1010 DATA 0.8, 0.3, 0.3
1020 DATA 0.1, 0.4, 0.3
1030 DATA 0.1, 0.3, 0.4
```

To compute the sales for next month from the sales for this month we need to carry out the matrix multiplication

$$\begin{bmatrix} 0.8 & 0.3 & 0.3 \\ 0.1 & 0.4 & 0.3 \\ 0.1 & 0.3 & 0.4 \end{bmatrix} \begin{bmatrix} 2000 \\ 3000 \\ 5000 \end{bmatrix}$$

where the present sales are 2000, 3000 and 5000 respectively. The resulting vector may in turn be multiplied by the matrix to give the sales for the month after next, and so on. Program 5.3 provides for this repeated multiplication of the vector by the matrix and so the long-term trend in sales may be investigated. Various starting vectors should be used to see how the system behaves in various cases.

The reader may already be aware that the eigenvectors and eigenvalues of the matrix play an important part in the above situation. For any matrix M the sequence of vectors

$$\mathbf{V} \quad M\mathbf{V} \quad M^2\mathbf{V} \quad M^3\mathbf{V} \quad M^4\mathbf{V} \ldots$$

tends towards a vector in the direction of the eigenvector with the largest eigenvalue. In the case of stochastic matrices, such as the example given above, the largest eigenvalue is 1 and the sequence of vectors tends towards a limiting vector. For other matrices there may be no limiting vector, in fact with largest eigenvalue greater than 1 the vectors increase in size. However, the direction of the vectors gets closer to that of the eignevector with largest eigenvalue. Since the vectors are increasing in size, a large number of multiplications by the matrix results in very large numbers and perhaps an overflow error. Program 5.3 may be used to find the direction of the eigenvector numerically. This calculation of the eigenvector is easier to handle if each vector is scaled down as soon as it is calculated. Provided that the first component of the eigenvector is non-zero then dividing through by this first component is a suitable scaling. The resulting vector is thus 'normalised' so that its first component is unity, and it is easy to see when convergence to the eigenvector has been achieved. We can also obtain the corresponding eigenvalue at the same time. Since for eigenvector $\mathbf{u}$ with eigenvalue α

$$M\mathbf{u} = \alpha\mathbf{u}$$

then each time the vector is multiplied by the matrix the component that was 1 becomes α. Thus the first component *before* scaling converges to the eigenvalue. To modify Program 5.3 to do this, all that is necessary is to change line 485 to do the scaling to

```
485  V(J) = X(J)/X(1)
```

We could also add

```
335  PRINT "EIGENVALUE ESTIMATE =";X(1)
```

to print out the eigenvalue.

In the case of 2×2 matrices each represents some transformation of the coordinate plane. If the computer has a graphics facility it adds insight to the student's explorations if the computer draws the vector after each multiplication by the matrix. Some of the techniques used in the transformation geometry section of the next chapter may be added into Program 5.3 to plot the vector. The experience of running the program with the vectors appearing one at a time provides a visual dynamic demonstration of the approach to the eigenvector direction. We suggest that students be encouraged to try the program with a variety of matrices and many different starting vectors for each. Discussion of the observed response can lead to the suggestion of a preferred direction for each matrix and paves the way for the idea of eigenvectors and eigenvalues. It is natural to ask what happens if we start with a vector in the special direction, or if there are any other directions that are special to the matrix. In suitably chosen cases this can lead to the discovery of an eigenvector whose eigenvalue is not the dominant one. The matrix

$$\begin{bmatrix} 1 & 2 \\ 2 & 1 \end{bmatrix}$$

has been found to be suitable for this since the second eigenvector is simple enough to be guessed.

5.3 TO MULTIPLY A MATRIX BY ANOTHER MATRIX

Implementations of BASIC on microcomputers do not usually include matrix operations, so if they are required then algorithms to perform them need to be written. Here we shall limit the discussion to the multiplication of two matrices. We have not found that we need to do this very often and so only brief mention is justified.

Consider the case where matrix A has to be multiplied by matrix B and the result stored in C. Naturally the dimensions need to conform so that if A is a $p \times q$ matrix and B is a $q \times r$ matrix then C must be $p \times r$.

$$\text{Row } J \rightarrow \underset{A}{\begin{bmatrix} \\ X\,X\,X\,X\,X\,X \\ \\ \end{bmatrix}} \underset{B}{\overset{\text{Column } K \downarrow}{\begin{bmatrix} X \\ X \\ X \\ X \\ X \\ X \end{bmatrix}}} = \underset{C}{\begin{bmatrix} \\ X \\ \\ \end{bmatrix}} \leftarrow \text{Element } c_{J,K}$$

The elements of row J of A are multiplied by the respective elements of column K of B and the results added to obtain element $c_{J,K}$. The counting along the row and down the column will be done by L. Each value of J from 1 to p must be combined with each value of K from 1 to r to obtain all the elements of C. A program to do all this is deceptively simple and we give an example in Program 5.4. Naturally for the result to be seen an output section will need to be included. Often the matrices are square and P, Q and R will all be the same and it is quite usual to use N for all of these throughout.

Program 5.4 – To multiply two matrices

```
100 REM**INPUT MATRICES**
105 READ P,Q,R
110 DIM A(P,Q), B(Q,R), C(P,R)
```

There follows sections to input the matrices A and B from the Data lines below. These two sections can be modelled on lines 130 to 150 of Program 5.1.

```
 200 REM**MULTIPLY MATRICES**
 210 FOR J = 1 TO P
 215    FOR K = 1 TO R
 220    C(J,K) = 0
 225       FOR L = 1 TO Q
 230          C(J,K) = C(J,K) + A(J,L)*B(L,K)
 235       NEXT L
 240    NEXT K
 245 NEXT J
1000 REM**DATA SECTION**
```

The values of P, Q and R and the elements of the matrices A and B can be put in DATA lines here and their order must be matched to the input sections.

5.4 TO INVERT A MATRIX

Unless it is quite small the inversion of a matrix involves a large amount of calculation. For 2 X 2 matrices the inverse can be written down at sight and students of elementary mathematics learn how to do so. Unfortunately the method does not extend to larger ones. The method that uses determinants requires their evaluation and besides being rather complex it is full of opportunities for errors.

There is a numerical method for the inversion of a matrix that is a development from the elimination used in the Gaussian Elimination method of solving simultaneous equations (see Chapter 13). We will start by describing the method and giving an example program, and then finally an explanation of why thc method works.

There are three stages in the process. (1) The matrix is reduced to upper triangular form by a succession of row operations in each of which a multiple of one row is added to another. The same operations are performed on another matrix that at the start of the process was the identity. (2) The upper triangular matrix is reduced to diagonal form by further row operations and the second matrix is treated similarly. (3) In the final stage each row is divided by the element on the leading diagonal (which is now the only non-zero element in the row) so that the matrix becomes the identity. The rows of the second matrix are also divided by the same value. This second matrix, that started out as the identity, will now have become the inverse.

In stage (1) the zeros are created one column at a time, and, as in Gaussian Elimination described in Chapter 13, we use K for the column where zeros are being created, and J for the row of the current zero. A multiple M of row K is added to row J to create the zero in $A(J, \mathrm{K})$. L is used to count through the elements of rows J and K as the addition takes place. The same value M is used to treat row J of matrix B in the same way. In Program 5.5 this part of the process is carried out in lines 300 to 360. The only variation from the corresponding section in the Gaussian Elimination program of Chapter 13 is the addition of lines 340 to 350 to treat matrix B in the same way as matrix A.

$$
\begin{array}{r}
\\ \\ \text{Row } K \rightarrow \\ \\ \text{Row } J \rightarrow \\ \\
\end{array}
\begin{array}{c}
\begin{array}{cccccc} & & \text{Column} & & & \\ & & K & & & \\ & & \downarrow & & & \end{array} \\
\left[\begin{array}{cccccc}
\ast & \ast & \ast & \ast & \ast & \ast \\
0 & \ast & \ast & \ast & \ast & \ast \\
0 & 0 & \ast & \ast & \ast & \ast \\
0 & 0 & 0 & \ast & \ast & \ast \\
0 & 0 & \ast & \ast & \ast & \ast \\
0 & 0 & \ast & \ast & \ast & \ast
\end{array}\right] \\
\text{Matrix } A
\end{array}
\begin{array}{c}
\left[\begin{array}{cccccc}
1 & 0 & 0 & 0 & 0 & 0 \\
\ast & 1 & 0 & 0 & 0 & 0 \\
\ast & \ast & 1 & 0 & 0 & 0 \\
\ast & \ast & \ast & 1 & 0 & 0 \\
\ast & \ast & 0 & 0 & 1 & 0 \\
\ast & \ast & 0 & 0 & 0 & 1
\end{array}\right] \\
\text{Matrix } B
\end{array}
$$

The state of the matrices just as the zero in $A(J,K)$ is about to be created, part-way through stage (1).
Non-zero elements are shown by *.

Stage (2) is carried out in a similar manner except that to create zeros above the leading diagonal we need to work backwards from the last column and proceed upwards in each column. Thus, in the program, lines 400 to 455 the K-loop counts the columns backwards from N to 2. For each value of K the rows are counted by J upwards from $K-1$ to 1. For each pair J, K a multiplier M is computed and L counts through the elements of the rows as M times row K is added to row J (lines 420 to 430). The same is done to the B matrix in lines 435 to 445.

$$
\begin{array}{r}
\\ \text{Row } J \rightarrow \\ \\ \\ \text{Row } K \rightarrow \\ \\
\end{array}
\begin{array}{c}
\begin{array}{cccccc} & & & & \text{Column} & \\ & & & & K & \\ & & & & \downarrow & \end{array} \\
\left[\begin{array}{cccccc}
\ast & \ast & \ast & \ast & \ast & 0 \\
0 & \ast & \ast & \ast & \ast & 0 \\
0 & 0 & \ast & \ast & 0 & 0 \\
0 & 0 & 0 & \ast & 0 & 0 \\
0 & 0 & 0 & 0 & \ast & 0 \\
0 & 0 & 0 & 0 & 0 & \ast
\end{array}\right] \\
\text{Matrix } A
\end{array}
\begin{array}{c}
\left[\begin{array}{cccccc}
\ast & \ast & \ast & \ast & \ast & \ast \\
\ast & \ast & \ast & \ast & \ast & \ast \\
\ast & \ast & \ast & \ast & \ast & \ast \\
\ast & \ast & \ast & \ast & \ast & \ast \\
\ast & \ast & \ast & \ast & \ast & \ast \\
\ast & \ast & \ast & \ast & \ast & \ast
\end{array}\right] \\
\text{Matrix } B
\end{array}
$$

The state of the matrices part-way through stage (2).
The zero in $A(J,K)$ is about to be created.

In the final stage in lines 500 to 535 each row of B is divided by the element on the leading diagonal of A. Line 530 is not really necessary but completes the process of converting A to the identity matrix.

$$\begin{bmatrix} * & 0 & 0 & 0 & 0 & 0 \\ 0 & * & 0 & 0 & 0 & 0 \\ 0 & 0 & * & 0 & 0 & 0 \\ 0 & 0 & 0 & * & 0 & 0 \\ 0 & 0 & 0 & 0 & * & 0 \\ 0 & 0 & 0 & 0 & 0 & * \end{bmatrix} \begin{bmatrix} * & * & * & * & * & * \\ * & * & * & * & * & * \\ * & * & * & * & * & * \\ * & * & * & * & * & * \\ * & * & * & * & * & * \\ * & * & * & * & * & * \end{bmatrix}$$

Matrix A Matrix B

The matrices at the end of stage (2).

The example given in the data lines of Program 5.5 is a 4 $\times$ 4 matrix that is reasonably easy to invert by hand. It is instructive for the pupil to check the stages from the computer. Small changes may easily be made to the program to output matrices A or B as required before the process is complete. For example, to see the state of A at the end of the first stage add the line

```
399 GOTO 900
```

and change 920 to

```
920 PRINT A(J,K);
```

Program 5.5 – To invert a matrix

```
100 REM**TO INVERT A MATRIX**
105 READ N
110 DIM A(N,N),B(N,N)
120 REM**INPUT MATRIX A**
125 FOR J = 1 TO N
130    FOR K = 1 TO N
135       READ A(J,K)
140    NEXT K
145 NEXT J
200 REM**SET MATRIX B=IDENTITY**
210 FOR J= 1 TO N
215    FOR K = 1 TO N
220       B(J,K)=0
225     NEXT K
230     B(J,J)=1
235 NEXT J
300 REM**REDUCE A TO UPPER TRIANGLE**
310 FOR K = 1 TO N-1
315    FOR J=K+1 TO N
320       M=-A(J,K)/A(K,K)
325       FOR L=K TO N
```

```
330  A(J,L)=A(J,L)+M*A(K,L)
335    NEXT L
340    FOR L=1 TO N
345      B(J,L)=B(J,L)+M*B(K,L)
350      NEXT L
355    NEXT J
360  NEXT K
400  REM**REDUCE TO DIAGONAL**
410  FOR K=N TO 2 STEP -1
415    FOR J=K-1 TO 1 STEP-1
420      M=-A(J,K)/A(K,K)
425      FOR L=K TO N
430        A(J,L)=A(J,L)+M*A(K,L)
435      NEXT L
440      FOR L=1 TO N
445        B(J,L)=B(J,L)+M*B(K,L)
450      NEXT L
455    NEXT J
460  NEXT K
500  REM**REDUCE TO IDENTIFY**
510  FOR J=1 TO N
515    FOR L=1 TO N
520      B(J,L)=B(J,L)/A(J,J)
525    NEXT L
530    A(J,J)=A(J,J)/A(J,J)
535  NEXT J
900  REM**OUTPUT INVERSE MATRIX**
910  FOR J=1 TO N
915    FOR K=1 TO N
920      PRINT B(J,K);
925    NEXT K
930    PRINT:PRINT
935  NEXT J
999  END
1000 REM  DATA SECTION
1005 DATA 4
1010 DATA 2,3,2,4
1020 DATA 4,9,6,12
1030 DATA 2,6,5,10
1040 DATA -2,3,4,12
```

We now return to the question of why the process works. Each step of the first two stages consists of adding a multiple M of row K to row J. This is equivalent to pre-multiplying A by a matrix such as

Column 2 ↓

$$\text{Row } 4 \rightarrow \begin{bmatrix} 1 & 0 & 0 & 0 & 0 & 0 \\ 0 & 1 & 0 & 0 & 0 & 0 \\ 0 & 0 & 1 & 0 & 0 & 0 \\ 0 & M & 0 & 1 & 0 & 0 \\ 0 & 0 & 0 & 0 & 1 & 0 \\ 0 & 0 & 0 & 0 & 0 & 1 \end{bmatrix}$$

In this example for a 6 X 6 matrix, M times row 2 is being added to row 4. We will call matrices like this 'elementary matrices'. For the last stage (stage (3)) the elementary matrices are of the form

$$\begin{bmatrix} 1 & 0 & 0 & 0 & 0 & 0 \\ 0 & 1 & 0 & 0 & 0 & 0 \\ 0 & 0 & 1 & 0 & 0 & 0 \\ 0 & 0 & 0 & \frac{1}{a_{44}} & 0 & 0 \\ 0 & 0 & 0 & 0 & 1 & 0 \\ 0 & 0 & 0 & 0 & 0 & 1 \end{bmatrix}$$

Let the elementary matrices for the steps taken in order be $E_1, E_2, E_3, E_4, \ldots, E_r$, then the process of reducing A to the identity may be written

$$E_r \ldots E_4 E_3 E_2 E_1 A = \mathrm{I}$$

so that the matrix product

$$E_r \ldots E_4 E_3 E_2 E_1$$

must give the inverse matrix for A. In the process described above we apply the same row operations to the matrix B that are applied to A. We thus compute $E_r \ldots E_4 E_3 E_2 E_1 B$ where B is the identity matrix. We will thus have found the inverse of A.

6

Computer geometry

The current generation of microcomputers with graphical displays, some in colour, provide an excellent vehicle for geometric exploration. Some of the techniques most appropriate for graphical displays, although not difficult, are frequently not included in mathematics syllabuses until the sixth form – if at all. It is possible, then, that this area of application of mathematics may be onc that has most influence on the content or ordering of future mathematics curricula. Without a doubt there is great satisfaction in being able to produce a program that generates a pleasing design on the screen and it is possible that this sense of pleasure and achievement can be tapped to provide good motivation for the study of geometry.

There are, though, some technical snags that make this route into geometry a little less straightforward than we would desire. To understand these we must make some assumptions about the way that the display screen is organised – these will vary from one model of microcomputer to another and we will assume a minimal basis for graphical displays. The precise commands used will vary and most manufacturers provide a wider range of facilities than assumed here. We assume, then, that (i) the graphics area of the screen is divided into a notional rectangular grid 320 divisions wide and 200 divisions high, (ii) that each little rectangle – known as a PIXEL – can be specified by a pair of co-ordinates relative to an origin at the bottom left-hand corner of the graphics area, (iii) that the progamming language used includes commands to select graphics, to fill a pixel and to draw a 'line' between specified pixels, and (iv) that we have only two colours to work with, that is, a background (usually black) and a foreground (often white or green).

6.1 POLYGONS, CIRCLES AND ELLIPSES

Within a programming language such as BASIC there is great diversity between manufacturers concerning the naming of graphics commands; we have chosen to use the names GRAPHICS, PLOT and LINE for our three elementary graphical instructions and Table 6.1 gives equivalents for some common microcomputers.

GRAPHICS	Has to be used before the PLOT and LINE commands can be obeyed. It clears the graphics area of the screen to the background colour.
PLOT X, Y	Colours in the pixel (X, Y) in the foreground colour.
LINE X, Y	Joins the last pixel used to the pixel (X, Y) by filling in a set of pixels that approximate the line joining the centres of the starting and finishing pixels – see Fig. 6.1.

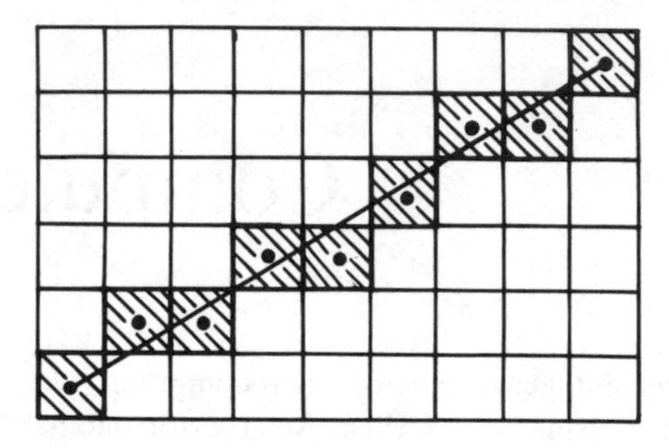

Fig. 6.1

Thus we can draw a figure such as a triangle by a simple program (see Program 6.1). Such an idea can, of course, be easily extended to produce general polygons.

Program 6.1 – To draw a triangle

```
100  REM PROGRAM 6.1 – A TRIANGLE
110  GRAPHICS
120  PLOT 10,10
130  LINE 310,10
140  LINE 160,190
150  LINE 10,10
```

Table 6.1

	GRAPHICS	PLOT X, Y	LINE X, Y
Apple	HGR : HCOLOR = 3	HPLOT X, Y	HPLOT TO X, Y
BBC	MODE 4	PLOT 69, X, Y	DRAW X, Y
380Z	CALL "RESOLUTION", 0, 2	CALL "PLOT", X, Y, 3	CALL "LINE", X, Y

The first and most serious snag that presents itself is that, in general, neither the individual pixels nor the graphics area of the screen is square. Thus the geometry of the microcomputer screen is akin to that of a piece of rectangular graph paper in which the gaps between lines in one direction are larger than the gaps in the other and in which the 'points' are represented by small rectangles. This would seem to make fundamental ideas such as line, angle, gradient and locus hard to interpret. What is more, the ratio of non-squareness of pixels can vary between different TVs and monitors connected to the same microcomputer. In order to determine this ratio a simple program is helpful (see Program 6.2).

Program 6.2 – To calibrate axes

```
100  REM PROGRAM 6.2 – TEST AXES
110  GRAPHICS
120  PLOT 110,100
130  LINE 210,100
140  PLOT 160,50
150  LINE 160,150
```

This program should produce a pair of lines in the centre of the screen each 100 units long. By measuring them with a ruler on the display screen you should be able to find the ratio between the horizontal and the vertical units. With luck these will be equal but on most systems the vertical unit is between 1.1 and 1.2 times the length of the horizontal unit. Our convention is to take the horizontal unit of length as the basic unit, called the SCREEN UNIT, and to introduce the idea of a 'squash factor', usually stored in a variable called *SF*, which is used to reduce the vertical units to the same scale as the horizontal units. We can thus make some simple modifications to Program 6.2 using Program 6.3. The particular value of *SF* will have to be adjusted to suit your particular system. It is worth emphasising that this program is applying some ideas from transformation geometry. First we move the 'origin' to the centre of the screen by the translation $\begin{bmatrix} 160 \\ 100 \end{bmatrix}$ and then apply a one-way stretch in the vertical direction to equalise the units of length. These ideas will be developed further throughout this chapter.

Program 6.3 – A pair of equal axes

```
100  REM PROGRAM 6.3 – EQUAL AXES
110  GRAPHICS
120  SF = 0.85
130  PLOT 110,100
140  LINE 210,100
150  PLOT 160,100 - 50*SF
160  LINE 160,100 + 50*SF
```

Now we have dealt with length along the axes we can consider the problem of drawing a line of given length (in screen units) and in a given direction. Consider, then, drawing a line 100 units long from the centre of the screen at an angle of 30° to the horizontal. The co-ordinates of the starting pixel A are (160,100) but which is the finishing pixel? In Fig. 6.2 AB is to be 100 units long and thus AP = 100 cos 30° and BP = 100 sin 30°. Now AP is measured in screen units already but BP is in vertical units which can be converted by use of the squash factor. Thus B is at (160 + 100 cos 30°, 100 + 100 sin 30° $*SF$). If SF = 0.85 as before then B is at (247, 142) to the nearest pixel. In writing programs it must be remembered that the SIN and COS functions usually require angles to be measured in radians. Some versions of programming languages contain functions to convert between degrees and radians but without these we can easily define a constant DR used for conversion from degrees to radians. $DR = \pi/180 \approx 0.0174533$.

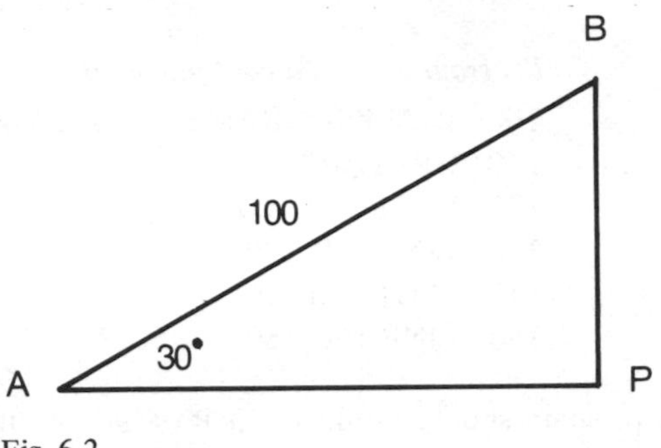

Fig. 6.2

6.1.1 Polygons

Suppose, then, that we wish to construct a hexagon in the centre of the screen. The size of a polygon could be specified by the length of a side or by the radius of the circle which circumscribes it. For a hexagon these two are identical and the angle subtended by each side at the centre is 60°.

Program 6.4 – To draw a polygon

```
100  REM PROGRAM 6.4 – A HEXAGON
110  GRAPHICS
120  H = 160 : K = 100 : R = 80 : SF = 0.85 : DR = 0.0174533
130  PLOT H + R, K
140  FOR AD = 60 TO 360 STEP 60
150      AR = AD*DR
160      LINE H + R*COS(AR), K + R*SIN(AR)*SF
170  NEXT AD
```

In Program 6.4 the centre is taken as (H,K) and the radius of the bounding circle as R. The hexagon will be traced out from the right-hand point in an anti-clockwise sense. The starting point has to be plotted and then each edge can be drawn by successive LINE commands. In order to close the polygon the last point used should be the same as the initial point, as in the triangle example of Program 6.1. To transform Program 6.4 to produce a regular polygon with N sides requires few changes. We must first specify the value of N and use it to calculate a step angle SA as in Fig. 6.3. As N may not divide 360 exactly it may be possible that the figure does not quite join up and so we must draw a final line to the starting point (Program 6.5).

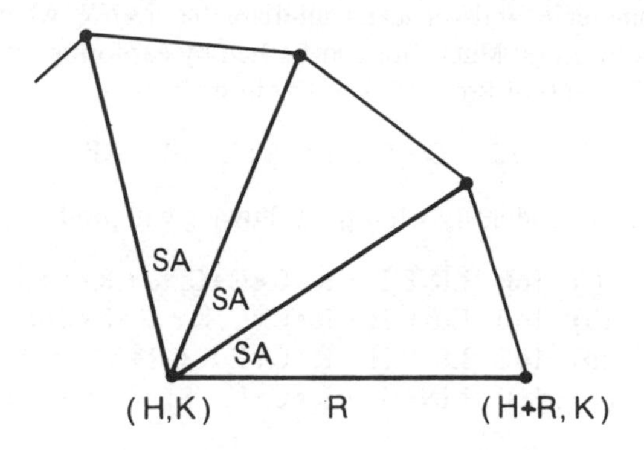

Fig. 6.3

Program 6.5 – To draw a regular polygon

```
100  REM PROGRAM 6.5 – AN N-GON
110  GRAPHICS
120  H=160 : K=100 : R=80 : N=11 : SF=0.85 : DR=0.0174533
125  SA = 360/N
130  PLOT H + R, K
140  FOR AD = SA TO 360 STEP SA
150     AR = AD*DR
160     LINE H + R*COS(AR), K + R*SIN(AR)*SF
170  NEXT AD
180  LINE H + R, K
```

6.1.2 Circles and ellipses

By using larger values for N we should get better approximations to a circle of radius R – in fact for values of N beyond 40 we do not gain much in accuracy but the figure takes longer to draw.

By changing the value of SF we can also see that the application of a one-way stretch to a circle produces an ellipse. The resulting ellipse has eccentricity e given by $e^2 = 1 - (b^2/a^2)$ so if we reduce SF by a half then $b = a/2$ and the ellipse has eccentricity $e = (\sqrt{3}/2)$. In fact the quantity b/a is probably a more useful measure of ellipticity and we shall call it ER, the ELLIPSE RATIO. Thus modifying line 160 of Program 6.5 to:

```
160  LINE H + R*COS(AR), K + R*SIN(AR)*ER*SF
```

and inserting a line such as:

```
122  ER = 0.5
```

we can control the ellipticity and so produce ellipses of any shape with principal axes parallel to the edges of the screen. In line 160 of Programs 6.4 and 6.5 we have the parametric equations of an ellipse centre (H,K) with horizontal semi-diameter R and vertical semi-diameter $R*ER$, where each length is measured in screen units. Much fun can be had by exploring other possible parametric forms in line 160 of Program 6.5. Try inserting:

```
155  C = COS(AR) : S = SIN(AR)
```

and try to identify what the following will produce:

```
(i)   160  LINE H + R*C*C*C, K + R*S*S*S*SF
(ii)  160  LINE H + R*C*C, K + R*S*S*SF
(iii) 160  LINE H + R*C↑P,K + R*S↑P*SF      for various P
(iv)  160  LINE H + R*C*(1 - S),K + R*S*(1 - C)*SF
```

You may have to adjust the starting and finishing points and (iii) may cause problems if you use a fractional or negative value of P when S or C become negative – see if you can think of cures to fix these bugs.

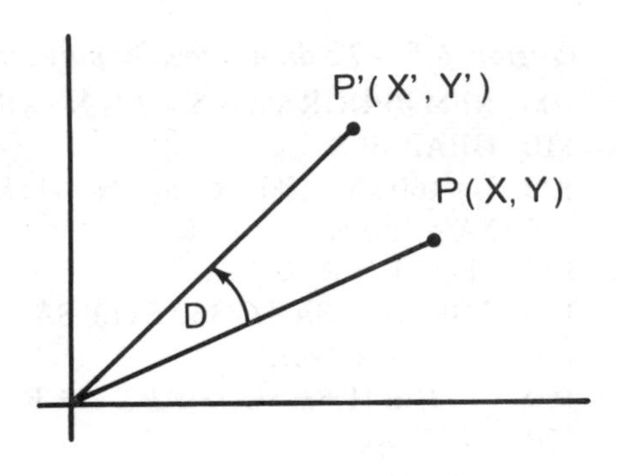

Fig. 6.4

An ellipse produced by the modified version of Program 6.5 will have its axes parallel to edges of the screen and so in order to place it in a general position we need to be able to perform a rotation about its centre. If a point $P(X,Y)$ transforms to the point $P'(X',Y')$ under an anti-clockwise rotation through angle D about the origin (see Fig. 6.4) then the relation between co-ordinates is given by

$$X' = X\cos D - Y\sin D$$
$$Y' = X\sin D + Y\cos D$$

which can also be expressed in matrix form as

$$\begin{pmatrix} X' \\ Y' \end{pmatrix} = \begin{pmatrix} \cos D & -\sin D \\ \sin D & \cos D \end{pmatrix} \begin{pmatrix} X \\ Y \end{pmatrix}$$

Thus to produce points on an ellipse of semi-diameters R and $R*ER$ rotated through an angle of D radians about its centre (H,K) we can consider the whole process as a sequence of transformations applied one after the other starting with an initial point $(\cos\theta, \sin\theta)$ which lies on the unit circle centred at the origin.

(i) E, an ENLARGEMENT, takes the unit circle to one of radius R

$$E: (\cos\theta, \sin\theta) \longrightarrow (R\cos\theta, R\sin\theta)$$

(ii) S_1, a ONE-WAY STRETCH, takes the circle of radius R into an ellipse

$$S_1: (R\cos\theta, R\sin\theta) \longrightarrow (R\cos\theta, R\sin\theta * ER)$$

(iii) R, a ROTATION, takes the ellipse of ellipse-ratio ER into a tilted ellipse

$$R: (R\cos\theta, R\sin\theta * ER) = (X, Y) \longrightarrow (X\cos D - Y\sin D, X\sin D + Y\cos D) = (X', Y')$$

(iv) S_2, another ONE-WAY STRETCH, compensates for the vertical screen scaling

$$S_2: (X', Y') \longrightarrow (X', Y' * SF)$$

(v) T, a TRANSLATION, moves the tilted and scaled ellipse to its correct centre

$$T: (X', Y' * SF) \longrightarrow (H + X', K + Y' * SF)$$

We can express this sequence in the form $\mathbf{p}' = TS_2RS_1E\mathbf{p}$ where $\mathbf{p}$ is the position vector of a point on the unit circle centre the origin and $\mathbf{p}'$ is the vector of the corresponding point on the tilted and scaled ellipse with shifted centre. The order of some of these transformations is quite critical. Performing R before S_1 would only have rotated the circle!

Program 6.6 – To draw a tilted ellipse

```
100 REM PROGRAM 6.6 – A TILTED ELLIPSE
110 GRAPHICS
120 H = 160 : K = 100 : R = 80 : N = 40 : SF = 0.85 :
    DR = 0.0174533 : ER = 0.5
125 SA = 360/N : DD = 30 : D = DD*DR : SD = SIN(D) :
    CD = COS(D)
130 PLOT H + R*CD, K + R*SD*SF
140 FOR AD = SA TO 360 STEP SA
150    AR = AD*DR
160    X = R*COS(AR) : Y = R*SIN(AR)*ER
170    XT = X*CD - Y*SD : YT = X*SD + Y*CD
180    LINE H + XT, K + YT*SF
190 NEXT AD
200 LINE H + R*CD, K + R*SD*SF
```

6.1.3 Efficiency

This is an appropriate moment to consider the issue of efficiency in writing programs. In Program 6.6, line 125, $\sin D$ and $\cos D$ are computed and then stored away as constants SD and CD to be used in line 170. If line 170 had been written as:

```
170 XT = X*COS(D) - Y*SIN(D) : YT = X*SIN(D) + Y*COS(D)
```

then cos D and sin D would have worked out (using some terms of a series) twice each for each pass of the AD loop, that is, 160 trigonometric evaluations would have been performed instead of the 2 in Program 6.6. This is a feature of an INTERPRETED language such as BASIC in which each line of the program is translated each time it is encountered – it has no 'memory' of having previously translated the same line. This is not the case in a COMPILED language such as FORTRAN, but the general point about keeping alert to possible inefficiencies in progamming is worth emphasising. If there is any doubt about whether a particular change, such as using A∗A in place of A↑2, will speed things up then it is easy to write a little test program. You can then time the program with a watch, alter line 130 to 130 B = A∗A and retime it.

Program 6.7 – To test the efficiency of squaring.

```
100  REM PROGRAM 6.7 – TEST OF EFFICIENCY
110  A = 27.48
120  FOR I = 1 TO 1000
130     B = A↑2
140  NEXT I
150  PRINT "DONE"
```

From this discussion we can test whether the use of a rotation transformation would speed up the process of drawing a 'circle'. Each new point is generated by rotating its predecessor by $SA = 360/N$ degrees where N is the number of points used in the polygonal approximation to the circle. It is useful to turn this program

Program 6.8 – To draw a circle quickly

```
100  REM PROGRAM 6.8 – A FASTER CIRCLE
110  H = 160 : K = 100 : R = 80 : N = 40 : SF = 0.85
120  A = 6.28318/N : S = SIN(A) : C = COS(A)
125  GRAPHICS
130  X = R : Y = 0 : PLOT H + R, K
140  FOR I = 1 TO N
150     XT = X∗C – Y∗S
160     Y = X∗S + Y∗C
170     X = XT
180     LINE H + X, K + Y∗SF
190  NEXT I
```

into a subroutine where the values of H, K and R are passed as parameters. If, however, it is required to generate many circles or ellipses then it is worth taking a different approach by storing tables of sines and cosines, that is, a set of points on the unit circle and to generate others by stretching, scaling and translating.

Program 6.9 – To draw many circles

```
 100 REM PROGRAM 6.9 – MANY CIRCLES
 110 DIM S(40),C(40)
 120 SF = 0.85 : SA = 6.28318/40 : S(0) = 0 : C(0) = 1
 125 GRAPHICS
 130 FOR I = 1 TO 40
 140    A = I*SA
 150    C(I) = COS(A) : S(I) = SIN(A)
 160 NEXT I
 200 H = 160 : K = 100
 210 FOR R = 10 TO 90 STEP 5
 220    GOSUB 1000
 230 NEXT R
 240 END
1000 REM – SUBROUTINE TO DRAW A CIRCLE
1010 YR = R*SF
1020 PLOT H + R, K
1030 FOR I = 1 TO 40
1040    LINE H + R*C(I), K + YR*S(I)
1050 NEXT I
1060 RETURN
```

Using some of the principles described in this section it should be possible to generate any curve from its parametric equations $x = f(t)$, $y = g(t)$ and to scale, rotate or translate it as required. With slight modifications it is also quite easy to produce curves from their polar equations $r = h(\theta)$. It is rather surprising that the form least suitable for computer display is the much more familiar cartesian equation. To plot a curve from an equation such as $x^2 + xy + y^2 = r^2$ would mean that an equation would have to be solved to find the values of y corresponding to each value of x, say. Furthermore this treatment of a curve by considering one variable to be dependent on the other may well destroy some of the inherent symmetries of the curve. In general, then, we could use the micro-computer display screen as a geometric test-bed for the exploration of new curves as well as the display of the familiar. In addition, since the projection of a circle is an ellipse, it is also possible to produce 3D impressions of objects such as pyramids and cones.

6.2 LINES AND VECTORS

It might now seem retrogressive to consider a way in which a straight line may be generated, especially since we have assumed the existence of a LINE command. However, we have only assumed a monochrome display with no intensity control and so for some display work will be useful to be able to draw a dotted or broken line. The relevant technique we shall use here is vector geometry and we shall start with the equation of a straight line (see Fig. 6.5). Suppose A and B have position vectors **a** and **b** relative to an origin O and that P is a point with position vector **p** somewhere on the line $\overrightarrow{AB}$. The vector AP, then, is in the same direction as $\vec{AB}$ and also is some multiple of it – let $\vec{AP} = t\vec{AB}$. If $t = 0$ then P is at A, if $t = 1$ then P is at B, if $0 < t < 1$ then P lies between A and B, if $t > 1$ then P lies outside B and if $t < 0$ then P lies outside A. Thus $\mathbf{p} = \vec{OP} = \vec{OA} + \vec{AP} = \vec{OA} + t\vec{AB}$ and $\vec{AB} = \vec{AO} + \vec{OB} = \vec{OB} - \vec{OA} = \mathbf{b} - \mathbf{a}$. Hence $\mathbf{p} = \mathbf{a} + t(\mathbf{b} - \mathbf{a})$, that is, $\mathbf{p} = (1 - t)\mathbf{a} + t\mathbf{b}$.

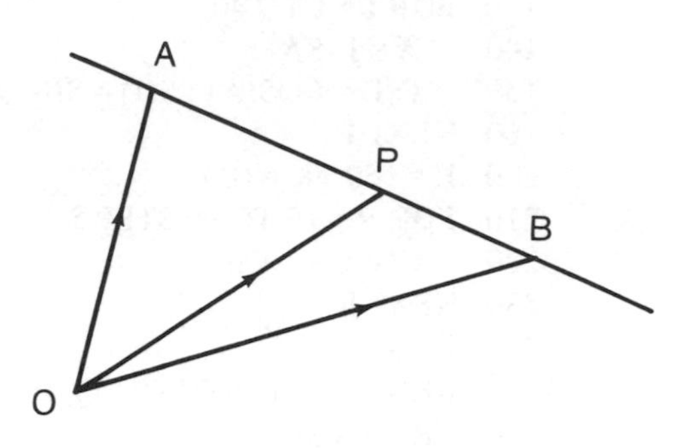

Fig. 6.5

6.2.1 Broken lines

Suppose, then, that O is the screen origin, A is the pixel (100, 50) and B is the pixel (200, 90). For any value of t we can generate a point P between A and B – for example if $t = 1/4$ then $P = (3/4)(100, 50) + (1/4)(200, 90) = (125, 60)$. Considering the differences in the x and y co-ordinates between A and B we see that the line from A to B has 100 pixels horizontally and 40 pixels vertically. As the horizontal distance is the greater it will be necessary to generate an appropriate pixel for every unit horizontal displacement from A in order for there to be no 'holes' in the line; if the vertical difference had been greater then we would need to generate a pixel for every vertical displacement. In our example if we take t increasing in steps of 1/100 from 0 to 1 and generate 101 pixels using $(1 - t)(100, 50) + t(200, 90)$ then we have the required line. If, however, we increased t by steps of 3/100 say, then the line will be dotted with only every third pixel displayed. This is not the most efficient way to generate

Progam 6.10 – To draw a dotted line

```
100  REM PROGRAM 6.10 – A DOTTED LINE
110  XA = 100 : YA = 50 : XB = 200 : YB = 90
120  DX = ABS(XA - XB) : DY = ABS(YA - YB)
125  GRAPHICS
130  D = DX : IF DY > DX THEN D = DY
135  REM-D IS NOW THE LARGER OF THE X AND Y
     DIFFERENCES
140  PLOT XA, YA
150  IF D < 3 THEN GOTO 220
160  S = 3/D
```

```
170 FOR T = S TO 1 STEP S
180    X = (1 - T)*XA + T*XB
190    Y = (1 - T)*YA + T*YB
200    PLOT X,Y
210 NEXT T
220 PLOT XB, YB
```

points on a line, though the program could easily be made much more efficient by replacing the multiplications in line 180 and 190 using the fact that the difference in x and y co-ordinates between neighbouring pixels are constant. We could modify Progam 6.10 by first finding the appropriate x and y increments XI and YI and then adding them successively. The appropriate modifications are as follows;

```
165 XI = S*(XB - XA) : YI = S*(YB - YA) : X = XA : Y = YA
170 FOR I = 1 TO D/3
180    X = X + XI
190    Y = Y + YI
210 NEXT I
```

6.2.2 Bezier curves

An important generalisation of the vector equation of the line is the basis of one of the major techniques for curve generation developed over the past 25 years. We can think of the vector equation of a line as being a WEIGHTED AVERAGE of the position vectors **a** and **b** where the WEIGHTS are linear functions of t. These weight functions have the property that each is non-negative for $0 < t < 1$ and that they add together to give 1 for any value of t. We could, then, think of generalising the idea to forming **p** as a weighted average of three vectors **a**, **b**, **c** where the weighting functions were quadratic in t, that is

$$\mathbf{p} = \alpha(t)\mathbf{a} + \beta(t)\mathbf{b} + \gamma(t)\mathbf{c} \ .$$

There are, of course, many possible choices for the weighting functions but a very profitable source is Pascal's triangle – or, more correctly, the binomial probability distribution. Part of the triangle of functions is shown in Fig. 6.6.

$$\begin{array}{ccccccccc}
 & & & & 1 & & & & \\
 & & & (1-t) & & t & & & \\
 & & (1-t)^2 & & 2t(1-t) & & t^2 & & \\
 & (1-t)^3 & & 3t(1-t)^2 & & 3t^2(1-t) & & t^3 & \\
(1-t)^4 & & 4t(1-t)^3 & & 6t^2(1-t)^2 & & 4t^3(1-t) & & t^4
\end{array}$$

Fig. 6.6

Thus we can investigate the kind of curve generated by three screen points, for example A(10, 10), B(100, 160), C(250, 50) as in Program 6.11.

Program 6.11 – To draw a Bezier curve

```
100  REM PROGRAM 6.11 – A BEZIER CURVE
105  DIM A(2), B(2), C(2), P(2)
110  A(1) = 10 : A(2) = 10 : B(1) = 100 : B(2) = 160 :
     C(1) = 250 : C(2) = 50
115  GRAPHICS
120  PLOT C(1), C(2) : PLOT B(1), B(2) : PLOT A(1), A(2)
130  FOR T = 0.05 TO 1 STEP 0.05
140    S = 1 - T : K1 = S*S : K2 = 2*S*T : K3 = T*T
150    FOR I = 1 TO 2
160      P(I) = K1*A(I) + K2*B(I) + K3*C(I)
170    NEXT I
180    LINE P(1), P(2)
190  NEXT T
200  LINE C(1), C(2)
```

The use of a list P(I) to store the co-ordinates of *P* enables the vector equation to be transformed into the loop in lines 150 to 170. This emphasises the point that the vector equation would apply in 3 (or more!) dimensions. The resulting curve should start from A tangential to AB and finish at C tangential to BC without passing through B. By holding points A and C fixed it is possible to change the shape of the curve by varying the position of B. An extension of the program to deal with 4 and 5 (or *N*) points allows quite complicated curves to be generated which can have inflections, loops and cusps. The class of such curves is known as Bezier curves after P. Bezier of the Renault car company who introduced them for use in interactive computer-aided design.

6.2.3 Clipping

Although many microcomputers now are capable of producing high-resolution graphic display, unfortunately not all have software to make this easy. In particular several microcomputers produce error messages when an attempt is made to plot a point or draw a line that extends past the edges of the screen. Thus it is very useful to have a subroutine which takes a pair of points $P_1(X1, Y1)$ and $P_2(X2, Y2)$ which are to be joined by a line and produces another pair of points $C_1(CX(1), CY(1))$ and $C_2(CX(2), CY(2))$ which are the edges of the intersection of the desired line with the screen. Such an intersection may take several forms – see Fig. 6.7.

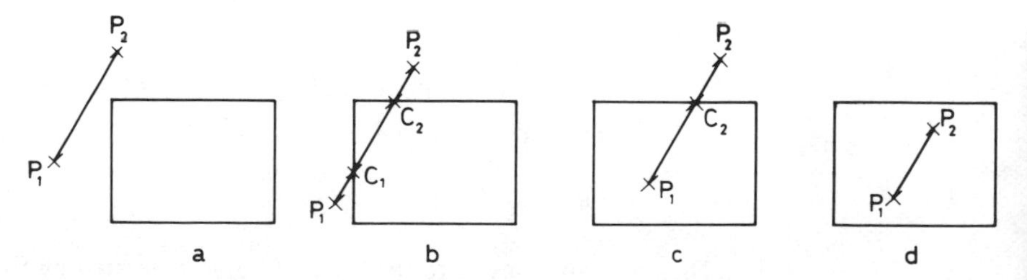

Fig. 6.7

First, then, we must determine whether the original line P_1P_2 has (i) both endpoints within the screen – in which case we can draw it straight away – or (ii) one endpoint within the screen – in which case we must find which of the four boundaries it intersects and determine the point of intersection – or (iii) both endpoints outside the screen – in which case we search for the pair of boundaries of the screen which it cuts, or else decide that it misses the screen altogether.

Suppose the four corners of the screen are (0,0), $(CW,0)$, (CW,CH) and $(0,CH)$ then the equations of the boundary lines are $X = 0, X = CW, Y = 0$ and $Y = CH$. In general, then, we wish to examine the intersections of the original line P_1P_2 with each of the four boundary lines. The vector equation of P_1P_2 is:

$$\begin{pmatrix} X \\ Y \end{pmatrix} = (1-t)\begin{pmatrix} X1 \\ Y1 \end{pmatrix} + t\begin{pmatrix} X2 \\ Y2 \end{pmatrix}$$

so to find its intersection with the left-hand edge of the screen $X = 0$ we have

$$\begin{pmatrix} 0 \\ Y \end{pmatrix} = (1-t)\begin{pmatrix} X1 \\ Y1 \end{pmatrix} + t\begin{pmatrix} X2 \\ Y2 \end{pmatrix} \tag{6.1}$$

that is,

$$(1-t)X1 + t.X2 = 0$$

so

$$t = \frac{X1}{X1-X2} \; .$$

However points on the line-segment P_1P_2 are given by values of t between 0 and 1 – thus P_1P_2 only intersects the left-hand edge of the screen between P_1 and P_2 if $0 \leqslant X1/(X1 - X2) \leqslant 1$. However, this intersection, if it exists, is not necessarily on the screen – substituting t in (6.1) gives $Y = Y1 + t(Y2 - Y1)$ which only lies at the edge of the screen if $0 \leqslant Y \leqslant YH$. Hence searching for an intersection of P_1P_2 with a boundary involves finding the intersection of two lines and then determining whether that point lies within both of the line segments as in Fig. 6.7(c). In writing the subroutine we take the four values of $X1$, $Y1$, $X2$ and $Y2$ as *input parameters* and adopt a convention that all variables that are *local* to the subroutine (see Program 6.12) will be given two letter identifiers with C (for 'clipping') as their first letter.

Program 6.12 – To draw the part of a line that falls within the screen

```
60000 REM PROGRAM 6.12 – TO CLIP THE LINE (X1,Y1) TO
      (X2,Y2)
60005 CW=319 : CH=199 : REM DIMENSION OF THE SCREEN
60010 C1 = -1 : C2 = -1 : REM VALUES FOR 'TRUE'
60020 IF X1<0 OR X1>CW OR Y1<0 OR Y1>CH THEN
      C1 = 0
60030 IF X2<0 OR X2>CW OR Y2<0 OR Y2>CH THEN
      C2 = 0
```

```
60040 IF C1 AND C2 THEN CX(1) =X1 : CY(1) = Y1 :
      CX(2) = X2 : CY(2) = Y2 : GOTO 61000
60050 IF (NOT C1) AND (NOT C2) THEN GOSUB 60080 :
      RETURN
60060 IF C1 THEN CI = X1 : CJ = Y1 : CO = X2 : CP = Y2 :
      GOSUB 60370 : RETURN
60070 IF C2 THEN CI = X2 : CJ = Y2 : CO = X1 : CP = Y1 :
      GOSUB 60370 : RETURN
60080 REM CLIP 2 POINTS
60090 CN = 0
60100 IF X1 = X2 THEN GOTO 60150
60110 CK = X1/(X1 - X2) : IF CK < 0 OR CK > 1 THEN GOTO
      60150
60120 CY = Y1 + CK*(Y2 - Y1)
60130 IF CY < 0 OR CY > CH THEN GOTO 60150
60140 CN = CN + 1 : CX(CN) = 0 : CY(CN) = CY
60150 IF Y1 = Y2 THEN GOTO 60210
60160 CK = Y1/(Y1 - Y2) : IF CK < 0 OR CK > 1 THEN GOTO
      60210
60170 CX = X1 + CK*(X2 - X1)
60180 IF CX < 0 OR CX > CW THEN GOTO 60210
60190 CN = CN + 1 : CX(CN) = CX : CY(CN) = 0
60200 IF CN = 2 THEN GOTO 60350
60210 IF X1 = X2 THEN GOTO 60280
60220 CK = (X1 - CW)/(X1 - X2)
60230 IF CK < 0 OR CK > 1 THEN GOTO 60280
60240 CY = Y1 + CK*(Y2 - Y1)
60250 IF CY < 0 OR CY > CH THEN GOTO 60280
60260 CN = CN + 1 : CX(CN) = CW : CY(CN) = CY
60270 IF CN = 2 THEN GOTO 60350
60280 IF Y1 = Y2 THEN RETURN
60290 CK = (Y1 - CH)/(Y1 - Y2)
60300 IF CK < 0 OR CK > 1 THEN RETURN
60310 CX = X1 + CK*(X2 - X1)
60320 IF CX < 0 OR CX > CW THEN RETURN
60330 CN = CN + 1
60340 CX(CN) = CX : CY(CN) = CH
60350 GOTO 61000
60360 RETURN
60370 REM CLIP 1 POINT
60375 CX(1) = CI : CY(1) = CJ
60380 IF CI = CO THEN GOTO 60420
60390 CK = CI/(CI - CO) : IF CK < 0 OR CK > 1 THEN GOTO
      60420
60400 CY = CJ + CK*(CP - CJ) : IF CY < 0 OR CY > CH THEN
      GOTO 60420
60410 CX(2) = 0 : CY(2) = CY : GOTO 61000
```

```
60420 IF CJ = CP THEN GOTO 60460
60430 CK = CJ/(CJ - CP) : IF CK < 0 OR CK > 1 THEN GOTO
      60460
60440 CX = CI + CK*(CO - CI) : IF CX < 0 OR CX > CW THEN
      GOTO 60460
60450 CX(2) = CX : CY(2) = 0 : GOTO 61000
60460 IF CI = CO THEN GOTO 60500
60470 CK = (CI - CW)/(CI - CO) : IF CK < 0 OR CK > 1 THEN
      GOTO 60500
60480 CY = CJ + CK*(CP - CJ) : IF CY < 0 OR CY > CH THEN
      GOTO 60500
60490 CX(2) = CW : CY(2) = CY : GOTO 61000
60500 IF CJ = CP THEN RETURN
60510 (CJ - CH)/(CJ - CP) : IF CK < 0 OR CK > 1 THEN
      RETURN
60520 CX = CI + CK*(CO - CI) : IF CX < 0 OR CX > CW THEN
      RETURN
60530 CX(2) = CX : CY(2) = CH
61000 PLOT CX(1), CY(1)
61010 LINE CX(2), CY(2)
61020 RETURN
```

Obviously it would be easy to replace the solid line drawn by lines 61000 and 61010 by a broken line by incorporating the techniques of Program 6.10. The technique can also be easily extended to enable drawing to take place within a rectangular sub-region of the display screen – known as a WINDOW.

6.3 3D REPRESENTATION AND MANIPULATION

So far we have looked at techniques for the generation of lines and curves in a plane but one of the most exciting uses of graphical displays is for representing three-dimensional objects. In order to describe a 3D object, such as a box, we need to specify some *geometrical* information, such as the co-ordinates of vertices, some *topological* information, such as a list of the vertices that are joined

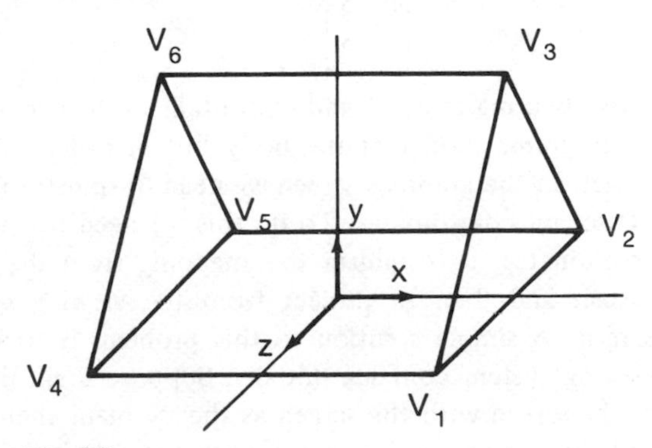

Fig. 6.8

by each edge and, possibly, some additional information concerning the display, such as the colours in which edges or faces are to drawn in. For the purposes of this example consider a simple shape such as a triangular prism, that is, a solid in the shape of a simple ridge tent or that of the container for a certain bar of chocolate shown in Fig. 6.8. Firstly we need to fix the axes in order to specify co-ordinate information.

6.3.1 Data structure

Taking the origin in the centre of the rectangular base of the prism with the x axis parallel to the long edges V_4V_1, V_5V_2, V_6V_3 and the z axis parallel to the short edges V_2V_1 and V_5V_4 then the y axis will be vertically upward like a central tent-pole in order to give a right-handed set of axes. Choosing some suitable units of length we can provide co-ordinates for each of the vertices of the body for example V_1 (100,0,40) and V_2 (100,0,−40). These co-ordinates can be stored in a matrix or array that we shall call C – thus for this example C is a 6 by 3 matrix.

$$C = \begin{bmatrix} 100 & 0 & 40 \\ 100 & 0 & -40 \\ 100 & 80 & 0 \\ -100 & 0 & 40 \\ -100 & 0 & -40 \\ -100 & 80 & 0 \end{bmatrix}$$

In the simplest kind of DATA STRUCTURE we need only define the edges of the body by giving the pair of vertices which each join – this information is held in an INCIDENCE MATRIX A. In our example there are 9 edges so A is 9 by 2.

$$A = \begin{bmatrix} 1 & 2 \\ 2 & 3 \\ 3 & 1 \\ 1 & 4 \\ 4 & 6 \\ 6 & 5 \\ 5 & 4 \\ 3 & 6 \\ 2 & 5 \end{bmatrix}$$

These two matrices, A and C, contain all the necessary information to construct a *wire-frame* model of the body but in order to display an impression of the object on the graphics screen we need to transform the 3D space co-ordinates to 2D screen co-ordinates. To do this we need to specify a system of 3D viewing co-ordinates, to establish the mapping from the body system to the viewing system and then to project from the viewing system onto the plane of the screen. A simple solution to this problem is to let the body system and the viewing system coincide initially. Suppose then, that the origin is at the centre of the screen with the screen as the xy plane then we can have the x axis across and the y axis up the screen with the z axis drawn out from the screen. The easiest kind of projection to implement is an orthogonal projection in which the mapping from a point in space (x, y, x) to its screen image (X, Y) is given simply by $X = x$ and $Y = y$. This is just equivalent to ignoring the z co-ordinate. The matrix of 3D co-ordinates C can be mapped under the projection into another matrix D called the DISPLAY FILE which holds the 2D screen co-ordinates ready for use in drawing the body. The display process simply entails for each edge finding its starting vertex S and finishing vertex F from the incidence matrix A, finding their 2D screen co-ordinates from the display file D and drawing a line on the screen from one to the other.

Program 6.13 – A simple 3D representation

```
 100 REM PROGRAM 6.13 – SIMPLE 3D DISPLAY
 110 REM – DEFINE A BODY
 120 V = 6 : E = 9
 130 DIM C(6,3), A(9,2), D(6,2)
 140 FOR I = 1 TO V
 150    READ C(I, 1), C(I, 2), C(I, 3)
 160 NEXT I
 170 DATA 100,0,40,100,0,-40,100,80,0
 180 DATA -100,0,40,-100,0,-40,-100,80,0
 190 FOR I = 1 TO E
 200    READ A(I, 1), A(I, 2)
 210 NEXT I
 220 DATA 1,2,2,3,3,1,1,4,4,6,6,5,5,4,3,6,2,5
 300 GOSUB 2000 : REM PERFORM PROJECTION
 400 GOSUB 3000 : REM DISPLAY BODY
 500 END
2000 REM – ORTHOGONAL PROJECTION
2010 FOR I = 1 TO V
2020    D(I, 1) = C(I, 1) + 160
2030    D(I, 2) = C(I, 2) + 100
2040 NEXT I
2050 RETURN
3000 REM – DISPLAY
3010 GRAPHICS
3020 FOR I = 1 TO E
3030    S = A(I, 1) : F = A(I, 2)
3040    PLOT D(S, 1), D(S, 2)
3050    LINE D(F, 1), D(F, 2)
3060 NEXT I
3070 RETURN
```

6.3.2 Projections

In our example (Program 6.13) this display would just produce a rectangle – the projection of the sloping face $V_1 V_3 V_6 V_4$ with all the other edges hidden beneath its sides. In Program 6.13 the matrix D is really redundant but it has been included so that the program can easily be modified to perform more complex projections.

For very complex bodies the matrices involved may become quite large and the program could become too big for some of the smaller microcomputers. Space savings may be possible – when matrices are dimensioned most systems count row and column numbers from 0 – thus our matrix *C* which was given dimension (6,3) would in fact have 7 rows and 4 columns. Space could thus be saved by using the 0, 1 and 2 columns instead of the 1, 2 and 3 columns for storing co-ordinates etc. Also some microcomputers store integers more efficiently than, so-called, floating point numbers and, as the information in the *A* matrix will always be integer, so that may afford savings. In addition we can only display points on the screen to the nearest pixel and so the information in the *D* matrix could be rounded to the nearest integer.

A projection we often use to represent solids while drawing on a board or overhead projector is an oblique projection. In this the *X* and *Y* co-ordinates of points away from the plane of projection $Z = 0$ are increased or decreased by amounts proportional to their *Z* co-ordinates – which is like performing a shear in three dimensions with $Z = 0$ as the invariant plane, followed by an orthogonal projection. This can be achieved by a few simple amendments to Progam 6.13 to change the projection used in setting up the display file *D*.

```
2005  M1 = 0.5 : M2 = 0.5
2020     D(I,1) = C(I,1) - M1*C(I,3) + 160
2030     D(I,2) = C(I,2) - M2*C(I,3) + 100
```

By experimenting with different values of *M*1 and *M*2 the obliqueness of the projection can be varied to any desired amount.

Another possible projection is point-perspective, or conical projection, which, surprisingly, is very simple to perform.

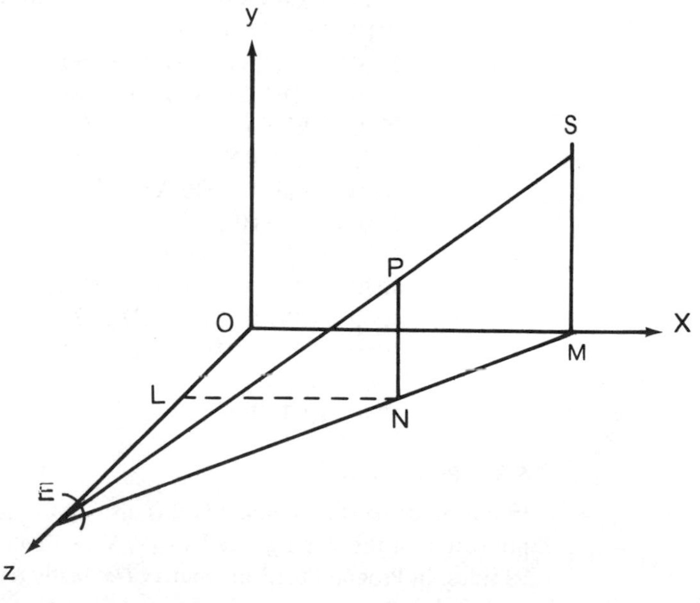

Fig. 6.9

In Fig. 6.9 P represents a point (x,y,z) in space and S its screen projection (X,Y) on the xy plane when viewed from a point $E(0,0,e)$ on the z axis. N and M are the feet of the perpendiculars from P and S to the xy plane and L is the foot of the perpendicular from N to the z axis. From the similar triangles ELN and EOM we have:

$$\frac{\mathrm{OM}}{\mathrm{LN}} = \frac{\mathrm{OE}}{\mathrm{LE}}$$

where LN = x, LE = $e-z$, OM = X and OE = e. Thus

$$\frac{X}{x} = \frac{e}{e-z}$$

so

$$X = x\frac{e}{e-z} \; .$$

Similarly, from the similar triangles EPN and ESM we have:

$$\frac{\mathrm{SM}}{\mathrm{PN}} = \frac{\mathrm{ME}}{\mathrm{NE}}$$

but, from triangles ELN and EOM:

$$\frac{\mathrm{ME}}{\mathrm{NE}} = \frac{\mathrm{OL}}{\mathrm{LE}} = \frac{e}{e-z} = \frac{Y}{y}$$

so

$$Y = y\frac{e}{e-z} \; .$$

Thus each point has its x and y co-ordinates multiplied by the factor $e/(e-z)$ which approaches 1 as the eye goes to infinity along the z axis, agreeing with orthogonal projection. Obviously there will be problems if $z = e$ as the point will not project onto the screen at all. Similarly if $z > e$ then the point lies behind the eye and should not be projected. We can incorporate this projection into a subroutine (see Program 6.14) to replace that for orthogonal projection.

Program 6.14 – A perspective projection

```
2000 REM PROGRAM 6.14 – PERSPECTIVE PROJECTION
2010 EZ = 600 : H = 160 : K = 100
2020 FOR I = 1 TO V
2030   Z = C(I,3)
2040   IF Z >= EZ THEN GOTO 2100
2050   M = EZ/(EZ - Z)
2060   D(I,1) = M*C(I,1) + H
2070   D(I,2) = M*C(I,2) + K
2080 NEXT I
2090 RETURN
2100 D(I,1) = 9999 : D(I,2) = 9999
2110 GOTO 2080
```

In order to avoid displaying edges which pass behind the eye we can insert

```
3035 IF D(S,1) = 9999 OR D(F,1) = 9999 THEN GOTO 3060
```

In the case of our triangular prism this projection should now look like a trapezium with triangles described at the ends.

6.3.3 Transformations

As a last technique it would be convenient to be able to move the body around in space by transformations such as rotations, translations, stretches, etc. For simplicity we can consider rotations about the viewing system axes. Suppose we wish to rotate the body about the x axis through an angle T then the x co-ordinates of the body will be unaltered and the rotation is just like a 2D rotation in the yz plane. Thus the transformation from (x,y,z) to the rotated co-ordinates (x',y',z') is given by:

$$\begin{aligned} x' &= x \\ y' &= y\cos T - z\sin T \\ z' &= y\sin T + z\cos T \end{aligned}$$

which can be represented by the matrix

$$\begin{bmatrix} 1 & 0 & 0 \\ 0 & \cos T & -\sin T \\ 0 & \sin T & \cos T \end{bmatrix}$$

It should be easy to derive the form of the other two principal rotations. We can then write such space transformations as subroutines (see Program 6.15).

Program 6.15 – A rotation in space

```
4000 REM PROGRAM 6.15 – ROTATE ABOUT X-AXIS
4005 REM THE VALUE OF T IS PASSED AS A PARAMETER
4010 S = SIN(T) : C = COS(T)
4020 FOR I = 1 TO V
4030    Y = C*C(I,2) - S*C(I,3)
4040    Z = S*C(I,2) + C*C(I,3)
4050    C(I,2) = Y : C(I,3) = Z
4060 NEXT I
4070 RETURN
```

Thus to rotate the body through 0.2 radians (about 12°) before projection we can link subroutine with the main program by inserting

```
250 T = 0.2 : GOSUB 4000
```

Now we should be able to see all the edges projected separately, We could then simply extend the 3D viewing program by adding a subroutine to perform each major type of transformation. For example a translation through a vector $(a\,b\,c)^T$ is specifed by the mapping:

$$x' = x + a \qquad y' = y + b \qquad z' = z + c\ .$$

In order to treat the body as a solid and to remove the hidden lines a more complex data structure would be needed that specified information about the faces of the body. Even then, for all but convex bodies, the elimination of hidden lines is a time-consuming process.

6.4 REFLECTIONS

With the quite small set of techniques presented here there is much scope for achieving some very attractive effects. It is, of course, quite possible to replicate diagrams from plane geometry showing triangles, bisectors, circumcircles and the like and, equally, to show the kind of representation of wedges etc. that were the stock of solid geometry. Similarly it is possible to produce 'animated' sequences to show properties of curves – such as generating a cycloid by displayed the position of a fixed point on each of a set of circles which appear to be rolling on a fixed line.

However, the great attraction of the classroom microcomputer lies in its capacity for interactive use and this is particularly appropriate to geometric exploration. By the use of INPUT statements, or the like, we can enable parameters of the program to be 'tweaked' by the user. For example it would be easy to make the parameter EZ that specifies the distance of the eye from the screen in the 3D program changeable by the user to see the effect of varying the perspective.

In the teaching of transformation geometry we have already provided many examples of the usefulness of the techniques but we can also produce simple programs that allow pupils to enter 2 by 2 or 3 by 3 matrices and to observe their geometric effect and so be able to experiment with the form of matrices, as well as experimenting with the order in which combinations of transformations are performed.

At a much earlier level a great deal can be discovered about angle and length and skills of estimation encouraged by the use of a simple drawing program where an outline is produced by 'driving a pen' around the screen controlled by simple commands such as FORWARD 50, LEFT 90, BACK 100 etc. These ideas are to be found in a progamming language designed for children called LOGO and also in certain programmable electronic toys such as the BIGTRAK tank.

It must be stressed that a program designed to give a feel for angle or length or 3D representation *must* be accurate and care must be taken over the problem of the non-square pixel – a program that is supposed to illustrate isometries and transforms a square into an oblong would be disastrous. Also we do not for one moment envisage the application of the microcomputer in geometry teaching as in any way a substitute for the other enjoyable skills such as using instruments for constructions or curve stitching, nor do we see its use as an exploratory instrument as in any way negating the need for proof. However, it is capable of performing many tasks that we cannot replicate in the classroom by other means and it does provide an admirable vehicle for forming and testing hypotheses and for stimulating discussion. In particular it adds a usefulness to geometric language – knowing what a translation is enables a child to have control over a program such as the 3D one and thus it can be used by the child in the way he or she wants.

We have made much use of such ideas as parametric equations, vectors and transformations and it may be that the power of such techniques in computer displays may have a bearing on the eventual emphasis we give to particular elements of the geometry syllabus. To finish with an optimistic anecdote: a mathematics teacher was recently confronted by a 12-year-old who had been working for some time on programming his own kind of computer battle game. Having seen some of the effects produced on the fairground and arcade machines he had decided that he wanted to be able to have spacecrafts spinning about on the screen but he did not know how to achieve the effect. Another child who regularly studied the hobby magazines on computing told him that he thought the technique involved the use of matrices. However, the 12-year-old had not yet been taught matrices so it seemed only natural to go to the maths teacher and demand to be taught matrices *now* because he thought he needed them!

7

Logic and computers

7.1 COMPUTER LOGIC

A computer is said to be a logical device as well as a number processor. It ought, therefore, to be possible to use it in the study of elementary logic, but how is this to be done? The BASIC language includes the logical connectives AND, OR and NOT but those who have tried to use them without understanding how they work, may have been surprised and puzzled by the results. By giving A and B various values we can investigate the effect of the command

PRINT A AND B

The result is a number that is related to the values given to A and B. Logical AND is associated with the intersection of two sets, and the computer works with numbers coded in binary form. Each binary digit (bit) may be thought of as an element in a set and AND produces the intersection of two sets. For example if A = 13 and B = 27 then PRINT A AND B produces 9 as the result.

	Base ten	Binary
A	13	0 1 1 0 1
B	27	1 1 0 1 1
A AND B	9	0 1 0 0 1

As can be seen from the binary representation, A AND B has produced a number that has a 1 in those positions where there are 1's in both A and B and 0's otherwise.

OR is associated with the union of sets and works in a similar way. It is the 'inclusive or' that corresponds to union and thus for example if A = 9 and B = 19 then A OR B produces the value 27.

	Base ten	Binary
A	9	0 1 0 0 1
B	19	1 0 0 1 1
A OR B	27	1 1 0 1 1

The result has a 1 in those positions where either A or B or both have 1's. Now that we understand the effect of AND and OR it should be possible to use the computer to handle logical processes. We will return to this point later, but one immediate application of the above is to test whether a number is even or odd. An even number has a 0 as its final digit whereas an odd number has 1, thus to print out whether A is even or odd we can use the program statements

```
IF (A AND 1) = 0 THEN PRINT "EVEN"
IF (A AND 1) = 1 THEN PRINT "ODD"
```

7.2 TRUTH TABLES

In the study of elementary logic, truth tables are used to show the value of a logical function for the various values of the variables. Conventionally we use 1 for true and 0 for false. An attempt to use these on the computer together with AND, OR and NOT producers some peculiar results. For reasons connected with the internal representation of numbers the computer uses –1 for 'true' and 0 for 'false'. It is thus possible to produce a truth table with these values, however it may be more convenient to change the sign when printing the result. An example is given in Program 7.1. Line 110 prints the headings for the table and to change to a different function only this and line 150 need to be changed.

Program 7.1 – To produce a truth table for the logical expression A AND (B OR C)

```
100 REM**TRUTH TABLE**
110 PRINT " A     B     C     A AND (B OR C)"
120 FOR A = 0 TO -1 STEP -1
130    FOR B = 0 TO -1 STEP -1
140       FOR C = 0 TO -1 STEP -1
150          PRINT -A, -B, -C, -(A AND(B OR C))
160       NEXT C
170    NEXT B
180 NEXT A
```

Output of Program 7.1

A	B	C	A AND (B OR C)
0	0	0	0
0	0	1	0
0	1	0	0
0	1	1	0
1	0	0	0
1	0	1	1
1	1	0	1
1	1	1	1

7.3 SETS AND SET LOGIC

We return to the idea of representing the membership of sets and the operations of complement, intersection and union. The basic idea is to make use of the way that the computer can treat the binary digits independently and let each set be represented by one of the binary digits. For example suppose we are classifying birds and the sets are {brown}, {small}, and {water} then three binary digits are needed for each bird. For example a curlew would be given the 'value' 101 to indicate that it is a brown, not small, water bird. Calling 101 the element's 'membership number', then, by knowing the membership number, we can tell the sets to which that element belongs. The problem is now one of defining the sets and allocating one of the binary places to each. The elements then need to be named and a membership number generated. This is similar to recording set membership on punched cards. Each element has a card and a series of holes near the edge are allocated to the sets. The sets to which the element belongs are indicated by converting the holes into slots. A stack of these cards may be sorted by putting a knitting needle through the hole corresponding to the desired set. The cards of all the elements belonging to that set fall off and are thus sorted from its complement.

We will now develop an example program in which the membership numbers for the elements are created and sorted by the computer in a way that is analogus to the punched cards. The following arrays will be used:

Names of the sets	S$
Binary digit indicating the set	S
Names of the elements	E$
Membership number of the element	E

For example, suppose the third set is {brown} then S$(3) will contain the word 'brown' and S(3) the binary number with 1 in position 3 and 0's everywhere else, that is, the number 100. Now if E$(7) is 'sparrow' then the membership number is E(7) will have a 1 in the third position to indicate that a sparrow is brown.

The first section of our Program 7.2 is concerned with naming the sets and allocating a binary digit to each. Lines 200 to 245 do this and in line 240 the number of sets is counted by N and the binary number for the set entered into

S(N). Since S(1) was set to 1 in line 220 the multiplication of the number in S(N – 1) by 2 results in a binary number with only one non-zero digit, and this in the next place along from the number for the set before. ZZZ is used to terminate the list and the test in line 235 only uses the first two Z's as a protection against the user accidentally typing the wrong number of Z's. Line 290 reduces N by 1 so that 'ZZZ' in the last store of array S$ will not be counted as a set name.

Now that the list of sets is complete the elements may be typed. As each one is named the user is asked to state to which sets it belongs, and this is like cutting the slots in the punched cards. In the program this is achieved by lines 300 to 390. The loop from lines 340 to 370 goes through the list of defined sets and asks if the element belongs to each. If it does then the input of a Y in response to line 350 causes the test in line 360 to allow control to pass on to line 365. The binary digit belonging to that set is then added to the membership number of the element. Again ZZZ is used to terminate the list.

The data base that has been created by these two sections may now be used in many ways. One is to offer a menu of options and ask the user to select what he wants the computer to do. In Program 7.2 the menu is offered by lines 400 to 490. Lines 500 to 585 offer the choice of a set whose members are then listed. The loop from line 550 to 565 examines each element in turn and at line 555 the logical AND tests for the occurrence of the binary digit of set C in the membership number of element J. If it is not present then control passes to line 565 and the element is not printed, whereas if it is present then the test statement is false and line 560 prints the element.

In a similar way lines 600 to 685 cause the complement of a set to be listed. This time the test in line 655 causes the printing of the element to be skipped if it is present in the set.

For the union of two sets the section in lines 700 to 797 tests each element at line 780. This time two similar tests are made and both must be 0 for the element to be omitted. This is achieved by the logical AND between the two tests. Intersection is similarly treated in lines 800 to 897 and it is line 880 that performs the vital test using logical OR.

Testing for membership of other set expressions could be achieved with these ideas but, depending on the complexity of the expressions, somewhat more elaborate programming may be needed. However, we have probably given enough indication in our example for the reader to develop the necessary program in accordance with his needs.

Program 7.2 – To create a data base of sets and elements and provide for listing of a set, a complement, a union or an intersection

```
100  REM**SET-LOGIC**
105  DIM S$(9), S(9), E$(50), E(50)
110  FOR J = 1 TO 50
115     E(J) = 0
120  NEXT J
200  REM**DEFINE SETS**
```

```
210 PRINT "NAME YOUR SETS"
211 PRINT "USE ZZZ TO INDICATE END OF LIST."
220 N = 1 : S(1) = 1
225 PRINT "DEFINE SET"; N
230 INPUT S$(N)
235 IF LEFT$(S$(N), 2) = "ZZ" THEN GOTO 290
240 N = N + 1 : S(N) = S(N - 1)*2
245 GOTO 225
290 N = N - 1
300 REM**INPUT ELEMENTS**
310 PRINT "NAME THE ELEMENTS"
311 PRINT "USE ZZZ TO END LIST."
312 PRINT "FOR EACH ELEMENT YOU WILL BE ASKED"
313 PRINT "TO WHICH SETS IT BELONGS."
314 PRINT "ANSWER Y OR N TO EACH."
320 M = 1
325 PRINT "WHAT IS ELEMENT"; M;
330 INPUT E$(M)
335 IF LEFT$(E$(M), 2) = "ZZ" THEN GOTO 390
340 FOR J = 1 TO N
345   PRINT "DOES IT BELONG TO {"; S$(J);"}"
350   INPUT T$
355   IF T$ = "N" THEN GOTO 370
360   IF T$ <> "Y" THEN GOTO 380
365   E(M) = E(M) + S(J)
370 NEXT J
375 M = M + 1 : GOTO 325
380 PRINT "YOU MUST TYPE Y OR N"
385 GOTO 350
390 M = M - 1
400 REM**OFFER OPTIONS**
410 PRINT "WOULD YOU LIKE TO :-"
411 PRINT "S - LIST A SET"
412 PRINT "C - LIST THE COMPLEMENT OF A SET"
413 PRINT "U - LIST THE UNION OF TWO SETS"
414 PRINT "I - LIST THE INTERSECTION OF TWO SETS"
415 PRINT "A - ADD MORE ELEMENTS TO YOUR LIST"
416 PRINT "E - END"
420 PRINT "TYPE LETTER OF OPTION REQUIRED."
425 INPUT T$
430 IF T$ = "S" THEN GOTO 500
435 IF T$ = "C" THEN GOTO 600
440 IF T$ = "U" THEN GOTO 700
445 IF T$ = "I" THEN GOTO 800
450 IF T$ = "A" THEN GOTO 325
455 IF T$ = "E" THEN GOTO 490
460 PRINT "YOU MUST TYPE A LETTER IN THE LIST."
```

```
465  GOTO 425
490  END
500  REM** LIST SET **
510  PRINT "LIST A SET. PLEASE TYPE THE NUMBER."
515  FOR J = 1 TO N
520    PRINT J;"{";S$(J);"}"
525  NEXT J
530  INPUT C$
535  C = VAL(C$)
540  IF C < 1 OR C > N THEN GOTO 580
545  PRINT "LIST OF {";S$(C);"}"
550  FOR J = 1 TO M
555    IF (S(C) AND E(J)) = 0 THEN GOTO 565
560    PRINT E$(J)
565  NEXT J
570  FOR K = 1 TO 1000 : NEXT K
575  GOTO 400
580  PRINT "YOU MUST CHOOSE A NUMBER"
581  PRINT "1 TO";N
585  GOTO 530
600  REM** LIST A COMPLEMENT **
610  PRINT "LIST COMPLEMENT. PLEASE TYPE NUMBER."
615  FOR J = 1 TO N
620    PRINT J;"{";S$(J);"}"
625  NEXT J
630  INPUT C$
635  C = VAL(C$)
640  IF C < 1 OR C > N THEN GOTO 680
645  PRINT "COMPLEMENT OF {";S$(C);"}"
650  FOR J = 1 TO M
655    IF (S(C) AND E(J)) <> 0 THEN GOTO 665
660    PRINT E$(J)
665  NEXT J
670  FOR K = 1 TO 1000 : NEXT K
675  GOTO 400
680  PRINT "YOU MUST CHOOSE A NUMBER"
681  PRINT "1 TO ";N
685  GOTO 630
700  REM**UNION OF TWO SETS**
710  PRINT "UNION REQUESTED. SELECT SETS FROM LIST."
715  FOR J = 1 TO N
720    PRINT J;"{";S$(J);"}"
725  NEXT J
730  PRINT "FIRST SET";
735  INPUT C$
740  C = VAL(C$)
745  IF C < 1 OR C > N THEN GOTO 730
```

```
750 PRINT "SECOND SET ";
755 INPUT D$
760 D = VAL(D$)
765 IF D < 1 OR D > N THEN GOTO 750
770 PRINT "UNION OF {";S$(C);"} AND {";S$(D);"}"
775 FOR J = 1 TO M
780   IF ((E(J) AND S(C)) = 0) AND ((E(J) AND S(D)) = 0)
      THEN GOTO 790
785   PRINT E$(J)
790 NEXT J
795 FOR K = 1 TO 1000 : NEXT K
797 GOTO 400
800 REM**INTERSECTION OF TWO SETS**
810 PRINT "INTERSECTION REQUESTED."
811 PRINT "SELECT SETS FROM LIST."
815 FOR J = 1 TO N
820   PRINT J;"{";S$(J);"}"
825 NEXT J
830 PRINT "FIRST SET ";
835 INPUT C$
840 C = VAL(C$)
845 IF C < 1 OR C > N THEN GOTO 830
850 PRINT "SECOND SET ";
855 INPUT D$
860 D = VAL(D$)
865 IF D < 1 OR D > N THEN GOTO 850
870 PRINT "INTERSECTION OF"
871 PRINT "{";S$(C);"} AND {";S$(D);"}"
875 FOR J = 1 TO M
880 IF ((E(J) AND S(C)) = 0) OR ((E(J) AND S(D)) = 0) THEN
    GOTO 890
885 PRINT E$(J)
890 NEXT J
895 FOR K = 1 TO 1000 : NEXT K
897 GOTO 400
```

8

Statistics

Statistics is an area of the curriculum where the use of a microcomputer seems obvious. The processing of long lists of data is just the sort of task that computers do well, but the way it is approached in hand calculations is often not the best on a computer. The calculation of the mean is a trivial one but indicates an approach to be used in calculating other statistics.

8.1 THE MEAN

For large values with relatively small dispersion the use of a false origin keeps the numbers in the calculation small and reduces the risk of errors in a hand calculation. A computer, working to say nine significant figures, does not produce significant errors and the use of a false origin is not necessary. In addition the false origin has to be chosen before any calculation begins and the whole of the data must be scanned in choosing a suitable value. With a computer the data values are used directly and part of the calculation can be done as the values are input. Often there is no need to store the data values but just accumulate various sums derived from them. The program will thus contain an input section in which the initial stages of the calculation are performed. The following simple program (Program 8.1) for the calculation of the mean illustrates this. The data items are typed in from the keyboard and the computer accumulates their sum and counts them. The end of the list is signalled by the user typing a 'rogue' value, in this case any negative number.

Program 8.1 – To calculate a mean

```
 90 REM**PROGRAM 8.1**
100 REM***CALCULATE MEAN***
110 PRINT "INPUT VALUES ONE AT A TIME."
115 PRINT "GIVE A NEGATIVE ROGUE VALUE TO END LIST."
120 N=1 : X1=0
125 REM***INPUT LOOP***
130 PRINT "VALUE";N;
135 INPUT X
140 IF X<0 THEN GOTO 200
145 X1=X1+X : N=N+1
150 GOTO 130
200 REM***CALCULATION OF MEAN***
210 N=N-1
215 MX=X1/N
220 PRINT "MEAN IS";MX
990 END
```

8.2 STANDARD DEVIATION

The variance of a set of numbers is defined as the average of the squares of the deviations from the mean. The standard deviation is the square root of the variance. We denote the set of numbers by x_i where i takes values from 1 to n, the mean by $\bar{x}$ and the standard deviation by σ_x. The formula for σ_x is then:

$$\sigma_x = \sqrt{\frac{\sum(x_i - \bar{x})^2}{n}} \tag{8.1}$$

where n is the number of values. To calculate σ_x directly from this formula the list must be scanned twice – once to calculate the mean and a second time to sum the squares of the deviations from the mean. Unless the individual numbers are stored this means that the list must by typed twice. In hand calculations this is overcome by working from a false origin in place of the mean and making a correction at the end. For computer calculation the false origin is most conveniently taken as 0. The new formula for σ_x is obtained as follows:

$$\sigma_x^2 = \frac{\sum(x_i - \bar{x})^2}{n} = \frac{\sum(x_i^2 - 2\bar{x}\,x_i + \bar{x}^2)}{n}$$

$$= \frac{\sum x_i^2}{n} - 2\bar{x}\frac{\sum x_i}{n} + \frac{n\bar{x}^2}{n}$$

$$= \frac{\sum x_i^2}{n} - 2\bar{x}\bar{x} + \bar{x}^2 \qquad \text{since} \quad \frac{\sum x_i}{n} = \bar{x}$$

$$= \frac{\sum x_i^2}{n} - \bar{x}^2$$

The formula (8.1) now becomes

$$\sigma_x = \sqrt{\frac{\sum x_i^2}{n} - \bar{x}^2} \qquad (8.2)$$

Program 8.1 may be simply modified to include the calculation of the standard deviation. The input loop needs to additionally accumulate the sum $\sum x_i^2$ as the data is entered. The calculation can then be done in a short additional section.

Program 8.2 – Additions to Program 8.1 for the calculation of the standard deviation

```
 90  REM**PROGRAM 8.2**
120  N = 1 : X1 = 0 : X2 = 0
145  X1 = X1 + X : X2 = X2 + X*X : N = N + 1
300  REM***STANDARD DEVIATION***
310  SX = SQR(X2/N - MX*MX)
315  PRINT "STANDARD DEVIATION IS"; SX
```

Variations of Programs 8.1 and 8.2 may be made to allow values to be read from a data section or input from a data file. A data section may be conveniently put at the end of the program from, say, line 1000. An advantage of using a data section is that it can be checked for errors and corrected without retyping the whole list. An alternative way of dealing with errors in input is to provide for the removal of the incorrect value. This would be a section of program to accept the incorrect value as input and subtract from stores X1, X2 and N. There would need to be some way of offering the user the selection of the correction section and this could be done by a particular value, say −1000. Care would of course need to be taken to see that this value is not used to end the list. In Programs 8.1 and 8.2 the line 137 IF X = -1000 THEN GOTO 900 would achieve this and the section at line 900 would need to end by passing control back to the input section at line 130 after the correction has been made.

The changes to Programs 8.1 and 8.2 to read the data from a data section are as follows:

Change the lines:

```
135  READ X
150  GOTO 135
```

Delete the lines 110, 115 and 130.

As an example of a data section

```
1000  DATA 34,36,37,45,43,46,42,48,38,39
1010  DATA 43,41,40,50,34,36,38,43,44,36
2000  DATA -100 : REM ROGUE TO END LIST
```

results in a mean of 40.65 and a standard deviation of 4.486. Insight into the behaviour of these two statistics can be given to students by running the program several times with different data sections. The data should be varied in many ways but one particularly instructive change is to introduce a value that is very different from the rest to see how one rogue value can affect the result. If the 36 in the above example is replaced by 136 then the mean changes to 45.65 as might be expected. The effect on the standard deviation is much more dramatic, however, and it increases to 21.80. It is instructive to discuss why this happens, and to relate it to the use of the squares of the deviations used in the calculation of the latter. It is also salutary to observe that the change of 36 to 136 could easily result from a simple typing error!

8.3 BIVARIATE DATA – PEARSON PRODUCT MOMENT CORRELATION

For each member of the population we have two different measurements x and y. For example, in an experiment to determine the effect of fertiliser the measurements might be the amount of fertiliser given and the crop weight. Thus x_5 would indicate the amount of fertiliser given to the fifth plant, and y_5 the weight of crop from that plant. If there is a relationship between the two then this may be indicated by a suitably computed statistic. Pearson's product moment correlation coefficient uses deviations from the mean. For a linear relationship $y = mx + c$ with positive m, both x and y values for a particular member of the population would be the same side of their respective means. Thus $(x - \bar{x})$ and $(y - \bar{y})$ will have the same sign and the product $(x - \bar{x})(y - \bar{y})$ will always be positive. If there is no relationship then for the population as a whole some products will be positive and some negative. Similarly with negative m in the relationship all products will be negative. Thus the summation $\sum(x_i - \bar{x})(y_i - \bar{y})$ will take large positive or negative values if there is a relationship, and near zero values if there is none. The actual value depends on the number in the population and the deviations of the x and the y values. It is therefore standardised by dividing by σ_x, σ_y and n to give Pearson's product moment correlation coefficient r where

$$r = \frac{\dfrac{\sum(x_i - \bar{x})(y_i - \bar{y})}{n}}{\sigma_x \sigma_y} \tag{8.3}$$

r can take values between 1 and -1. Of course exact relationships are seldom met in practice so with real data, values near 1 or -1 are significant and in any particular case the significance will need to be tested by a suitable table.

As it stands, equation (8.3) is not suitable for computer evaluation since for the summation $\sum(x_i - \bar{x})(y_i - \bar{y})$ the values of $\bar{x}$ and $\bar{y}$ would need to be available as the data is scanned. The summation may be rearranged as follows:

$$\frac{\sum(x_i-\bar{x})(y_i-\bar{y})}{n} = \frac{\sum x_i y_i}{n} - \bar{x}\frac{\sum y_i}{n} - \bar{y}\frac{\sum x_i}{n} + \frac{nxy}{n}$$

$$= \frac{\sum x_i y_i}{n} - \bar{x}\,\bar{y} - \bar{x}\,\bar{y} + \bar{x}\,\bar{y}$$

$$= \frac{\sum x_i y_i}{n} - \bar{x}\,\bar{y}\ .$$

Using this result (8.3) becomes:

$$r = \frac{\dfrac{\sum x_i y_i}{n} - \bar{x}\,\bar{y}}{\sigma_x\,\sigma_y}\ . \tag{8.4}$$

To calculate $\bar{x}$ and $\bar{y}$ we need $\sum x_i$ and $\sum y_i$ respectively. For σ_x and σ_y $\sum x_i^2$ and $\sum y_i^2$ are needed together with $\bar{x}$ and $\bar{y}$. The only other value required for the evaluation of (8.4) is the summation $\sum x_i y_i$. The program can easily be made to accumulate the five summations as the data is input or read. The values must of course be input as pairs x_i, y_i. In the Program 8.3 the data is input from the keyboard as a series of x, y pairs.

Program 8.3 – To calculate Pearson's product moment correlation coefficient

```
 90 REM**PROGRAM 8.3 **
100 REM**PROGRAM PEARSON CORRELATION**
110 PRINT "INPUT PAIRS OF VALUES X, Y"
115 PRINT "USE A NEGATIVE ROGUE PAIR TO END LIST"
120 N = 1 : X1 = 0 : Y1 = 0 : X2 = 0 : Y2 = 0 : XY = 0
125 REM***INPUT LOOP***
130 PRINT "X(";N;"),Y(";N;")";
135 INPUT X, Y
140 IF X < 0 THEN GOTO 200
145 X1 = X1 + X : Y1 = Y1 + Y : X2 = X2 + X*X :
    Y2 = Y2 + Y*Y : XY = XY + X*Y
150 N = N + 1 : GOTO 130
200 REM***CALCULATE STATISTICS***
205 N = N - 1
210 REM***MEANS***
215 MX = X1/N : MY = Y1/N
220 PRINT "MEAN X IS ";MX
225 PRINT "MEAN Y IS ";MY
```

```
300 REM***STANDARD DEVIATIONS***
310 SX = SQR(X2/N - MX*MX)
315 SY = SQR(Y2/N - MY*MY)
320 PRINT "ST. DEV. OF X IS ";SX
325 PRINT "ST. DEV. OF Y IS ";SY
400 REM***PEARSON CORRELATION***
410 RP = (XY/N - MX*MY)/(SX*SY)
420 PRINT "PEARSON CORRELATION IS "; RP
999 END
```

As with the previous programs, to gain the best experience from Program 8.3 it should be tried with a variety of data. Again it may be more convenient to use the data in the data section and the changes necessary are as follows.

```
Change:    135  READ X, Y
           150  N = N + 1 : GOTO 135
```

Delete lines: 110, 115 and 130.

As an example of a data section we used

```
1000 DATA 32,67,45,75,33,71,48,76,56,88
1010 DATA 28,61,33,62,46,78,48,80,38,67
1020 DATA 56,84,30,62,34,64,35,63,44,73
2000 DATA -100, -100 : REM ROGUE VALUES.
```

The data is in fifteen pairs with x before the corresponding y in each pair. Visual inspection shows that y is roughly 30 more than x in each case so we expect a positive correlation. The results are:

Mean of x is 40.4
Mean of y is 71.4
Standard deviation of x is 8.890
Standard deviation of y is 8.317

The Pearson product moment correlation coefficient is 0.951 and this figure confirms our expectations. A change of the 44 in the final pair to 144 has a dramatic effect. As observed before, the mean of x is not too greatly affected and becomes 47.1 but the standard deviation becomes 27.37. The effect on the correlation coefficient is equally dramatic and it falls to 0.356. There is no doubt that such numerical experimentation is valuable in getting a feel for these statistics.

8.4 REGRESSION LINE

Once it has been established that there is a significant correlation between x and y there arises the question of which linear equation best represents that relationship. To see how this relationship may be found consider the vary simple situation represented by the set of five values given in Fig. 8.1. A graph of this distribution suggests a relationship something like the line shown in Fig. 8.2. Consider the distances of the points above or below the line and marked d_i for $i = 1$ to 5.

x	2	3	6	7	8
y	3	5	6	7	6

Fig. 8.1.

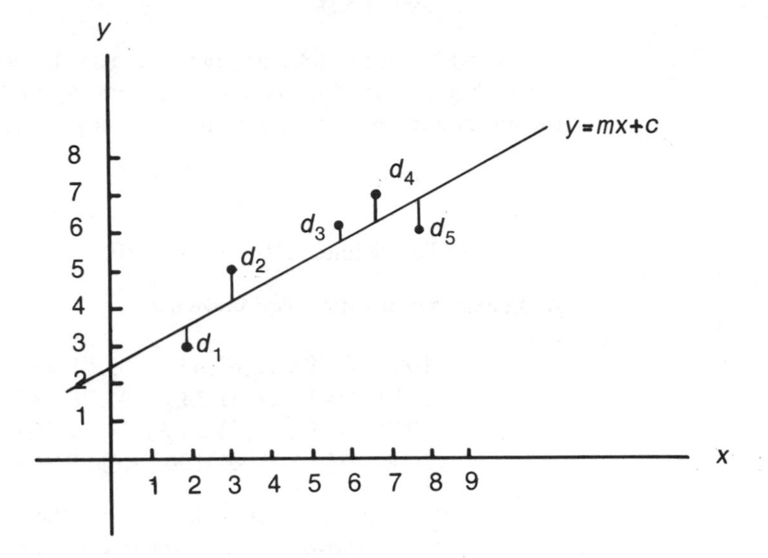

Fig. 8.2

The 'best' line can be taken to be the one that minimises the sum of the ds. Since positive and negative ds tend to cancel out we square them first to prevent this happening. We thus seek the line $y = mx + c$ that makes the value of $\sum d_i^2$ as small as possible. It may be shown that the required values of m and c are given by:

$$m = \frac{n\sum x_i y_i - \sum x_i \sum y_i}{n\sum x_i^2 - \left(\sum x_i\right)^2} \quad \text{or} \quad \left(\frac{\dfrac{\sum x_i y_i}{n} - \bar{x}\,\bar{y}}{\sigma_x^2}\right) \tag{8.5}$$

$$c = \frac{\sum x_i^2 \sum y_i - \sum x_i y_i \sum x_i}{n\sum x_i^2 - \left(\sum x_i\right)^2} \quad \text{or} \quad \left(\frac{\dfrac{\sum x_i^2}{n}\bar{y} - \dfrac{\sum x_i y_i}{n}\bar{x}}{\sigma_x^2}\right) \tag{8.6}$$

the computation of the parameters m and c of the regression line thus uses the same summations as those accumulated for Pearson's product moment correlation coefficient.

For our example distribution, Fig. 8.1, the summations are:

$$\sum x_i = 26,\ \sum y_i = 27,\ \sum x_i^2 = 162,\ \sum y_i^2 = 155,\ \sum x_i y_i = 154$$

and these values give $m = 0.507$ and $c = 2.761$. Comparison of these values with the line drawn visually in Fig. 8.2 gives confidence in the formulae.

The computation of the values of m and c for the regression line may be done by adding the following section to Program 8.3.

Program 8.4.

```
 90  REM**PROGRAM 8.4**
500  REM***REGRESSION Y ON X***
510  RM = (N*XY - X1*Y1)/(N*X2 - X1*X1)
515  RC = (X2*Y1 - XY*X1)/(N*X2 - X1*X1)
520  PRINT "REGRESSION LINE Y ON X IS"
525  PRINT "Y = M*X + C    WHERE"
530  PRINT "M = "; RM
535  PRINT "C = "; RC
```

The changes suggested on p. 105 can also be applied to Program 8.4. With the same data as suggested there we first find $m = 0.89$ and $c = 35.5$ so the line is $y = 0.89x + 35.5$ which is not very far from our rough observed value of $y = x + 30$. With the change of x from 44 to 144 in the last pair only, we have $m = 0.11$ and $c = 66.3$ which is a totally different line. This shows how just one false point can affect the results. Once again it is worthwhile experimenting with a variety of sets of data in the program.

8.5 SPEARMAN'S RANK CORRELATION COEFFICIENT

A popular non-parametric measure of correlation is Spearman's rank correlation coefficient. For this the data is first ordered by x-values and a rank r_{x_i} allocated to each. Tied values are given shared ranks. The list is then re-ordered by y-values and a rank r_{y_i} allocated to each. With perfect correlation the two ranks will be the same for each pair and the difference $r_{x_i} - r_{y_i}$ will always be zero. For negative correlation some differences will be positive and some negative. With the usual squaring to prevent postive and negative differences cancelling out, the summation

$$\sum (r_{x_i} - r_{y_i})^2 = \sum d_i^2$$

is zero for perfect positive correlation and takes its maximum value for perfect negative correlation. The size of the summation depends on the number of x,y pairs in the set and can be shown to have a maximum value of $n(n^2 - 1)/3$.

Thus the statistic

$$r_s = 1 - \frac{6 \sum d_i^2}{n(n^2 - 1)} \tag{8.7}$$

takes values between 1 and −1 and is Spearman's rank correlation coefficient.

As an example of the calculation of Spearman's rank correlation coefficient we consider again the very small set of values of Fig. 8.1. The ranking and calculation are then as follows.

x_i	r_x	y_i	r_y	d_i	d_i^2
2	5	3	5	0	0
3	4	5	4	0	0
6	3	6	2.5	0.5	0.25
7	2	7	1	1	1
8	1	6	2.5	−1.5	2.25

$$\sum d_i^2 = 3.5$$

$$r_s = 1 - \frac{6 \times 3.5}{5(25-1)} = 0.825 \ .$$

To compute a Spearman's rank correlation coefficient the whole list of data must be available and is therefore stored in an array to be available for sorting. The following example program uses the $n \times 4$ array W to hold the x, y pairs and the corresponding ranks. The program provides for the data to be input first and then the sorting is done in a subroutine. The ith row of the array holds the pair (x_i, y_i) and the corresponding ranks. The first column is for x values, the second for x ranks, the third for y values and the last for y ranks. Thus $W(I,1) = x_i, W(I,2) = r_{x_i}$, $W(I,3) = y_i$ and $W(I,4) = r_{y_i}$. The store S is set to either 1 or 3 according to whether x or y values are to be ranked. Even on a computer the sorting can take some time unless the list is quite a short one.

Program 8.5

```
 90 REM**PROGRAM 8.5 **
100 REM**SPEARMAN'S RANK CORRELATION**
105 PRINT "HOW MANY X,Y PAIRS";
110 INPUT N
115 DIM W(N,4)
120 REM***INPUT LOOP***
125 PRINT "INPUT PAIRS OF VALUES X,Y"
130 FOR J = 1 TO N
135 PRINT "X(";J;"),Y(";J;")";
140 INPUT W(J,1), W(J,3)
145 NEXT J
200 REM***RANK X-VALUES***
```

```
210  S = 1 : GOSUB 800
220  REM*** RANK Y-VALUES***
230  S = 3 : GOSUB 800
300  REM***CALCULATE SPEARMAN CORRELATION***
305  D2 = 0
310  FOR J = 1 TO N
315  D2 = D2 + (W(J, 2) - W(J, 4))↑2
320  NEXT J
330  RS = 1 - (6*D2)/(N*(N*N-1))
340  PRINT "SPEARMAN CORRELATION IS "; RS
350  END
800  REM***SORT SUBROUTINE***
810  F = 0
820  FOR J = 1 TO N - 1
825  IF W(J,S) >= W(J + 1,S) THEN GOTO 850
830  T1 = W(J, 1) : T2 = W(J, 2) : T3 = W(J, 3)
835  W(J, 1) = W(J + 1, 1) : W(J, 2) = W(J + 1, 2) : W(J, 3) =
     W(J + 1, 3)
840  W(J + 1, 1) = T1 : W(J + 1, 2) = T2 : W(J + 1, 3) = T3
845  F = 1
850  NEXT J
855  IF F = 1 THEN GOTO 810
860  REM***ALLOCATE RANKS***
865  J = 0
870  ER = 1
875  J = J + 1
880  IF J = N THEN GOTO 900
885  IF W(J, S) <> W(J + 1, S) THEN GOTO 900
890  ER = ER + 1
895  GOTO 875
900  FOR K = J - ER + 1 TO J
905  W(K, S + 1) = J - (ER - 1)/2
910  NEXT K
915  IF J < N THEN GOTO 870
920  RETURN
```

Some microcomputers, such as the BBC, require the array W to be filled before it can be used. This may be achieved by the program lines:

```
117  FOR J = 1 TO N
118    W(J, 2) = 0
119  NEXT J
```

Program 8.5 too may conveniently be modified to have a data section.

```
Change:  110  READ N
         135  READ W(J, 1), W(J, 3)
```

Delete the lines 105, 125 and 140.

The data section this time starts with the value for N. An example is

```
1000 DATA 15 : REM NUMBER OF X, Y PAIRS
1010 DATA 32,67,45,75,33,71,48,76,56,88
1020 DATA 28,61,33,62,46,78,48,80,38,67
1030 DATA 56,84,30,62,34,64,35,63,44,73
```

which is the same data that we used for Pearson's correlation coefficient. Spearman's coefficient calculated from this data is 0.92. This time the change of 44 to 144 in the last x-value does not produce such a dramatic effect and the coefficient falls to 0.84. Much useful discussion can result from trying a few examples.

8.6 BINOMIAL PROBABILITY DISTRIBUTION

In the early stages of learning the binomial probability distribution most of us would probably use Pascal's triangle for the binomial coefficients in evaluating the terms of the expansion of $(q+p)^n$. For small values of n and easy values of p a computer is hardly necessary – in fact its use may well detract from an understanding of the theoretical basis. There comes a point, however, where experience of large values of n and a wide range of values of p is desirable, and the calculation of many terms becomes tedious even with the aid of a calculator. Not only can a microcomputer easily compute the terms of the expansion but the VDU can be used to display the result as a histogram. This is especially useful when discussing the approach to the normal distribution as n increases. Hand calculation can begin to show this fairly convincingly when p and q are similar in size, but not so easily when p is small and n must become very large before the distribution looks normal. With a suitable computer program many combinations of p and n can be displayed in quick succession and we thus have a flexible tool with which to develop this important idea.

We need to evaluate the individual terms of the expansion:

$$(q+p)^n = q^n + {}^nC_1 q^{n-1}p + {}^nC_2 q^{n-2}p^2 + {}^nC_3 q^{n-3}p^3 + {}^nC_4 q^{n-4}p^4 + \dots {}^nC_r q^{n-r}p^r + \dots + p^n \qquad (8.8)$$

in which the ${}^nC_r$ are the binomial coefficients (sometimes written $\binom{n}{r}$). A formula for these coefficients is:

$${}^nC_r = \frac{n!}{r!\,(n-r)!} \qquad \text{for} \quad r = 1 \text{ to } n-1$$

and the probability of r successes in n trials is given by the term ${}^nC_r q^{n-r}p^r$ from the expansion.

For computer evaluation it is best to cancel the $(n-r)!$ with terms from $n!$.

The expansion may then be written:

$$(q+p)^n = q^n + \frac{n}{1}q^{n-1}p + \frac{n(n-1)}{1\times 2}q^{n-2}p^2 + + \frac{n(n-1)(n-2)}{1\times 2\times 3}q^{n-3}p^3 + \ \dots + \frac{n(n-1)(n-2)(n-3)\dots(n-r+1)}{1\times 2\times 3\times \dots\times r}q^{n-r}p^r + \dots + p^n \tag{8.9}$$

in which each term may be obtained from the previous term by multiplying by the value of $(n-r+1)$ and p, and dividing by r and q. The probability of no successes $P(0)$ is first computed and is q^n. We can then use the relation $P(r+1) = P(r) \times [(n-r+1)/r] \times p/q$ to compute each probability from the one before. In the following example program (Program 8.6) the array P is used for the probabilities and the store M for the value of $n-r+1$ at each stage. The relationship between probabilites is used in line 140.

Program 8.6 – To calculate a binomial probability distribution.

```
 90  REM**PROGRAM 8.6**
100  REM***BINOMIAL DISTRIBUTION***
110  PRINT "BINOMIAL PROBABILITY DISTRIBUTION"
115  PRINT "WHAT VALUES FOR N AND P";
120  INPUT N,P
125  DIM P(N)
130  Q = 1 - P : P(0) = Q↑N : M = N
135  FOR R = 1 TO N
140  P(R) = P(R - 1)*M/R*P/Q
145  M = M - 1
150  NEXT R
200  REM***OUTPUT LOOP***
210  FOR R = 0 TO N
215  PRINT P(R)
220  NEXT R
```

With large n we often only require part of the distribution and the program can easily be modified to achieve this. If n is large and p is small all but a few terms are negligible. We use FR and LR for the first and last values of r that are required and change the following lines in Program 8.6:

```
135  FOR R = 1 TO LR
210  FOR R = FR TO LR
```

We also need to input the range required and this may be done with:

```
122  PRINT "WHAT FIRST AND LAST VALUES OF R";
123  INPUT FR, LR
```

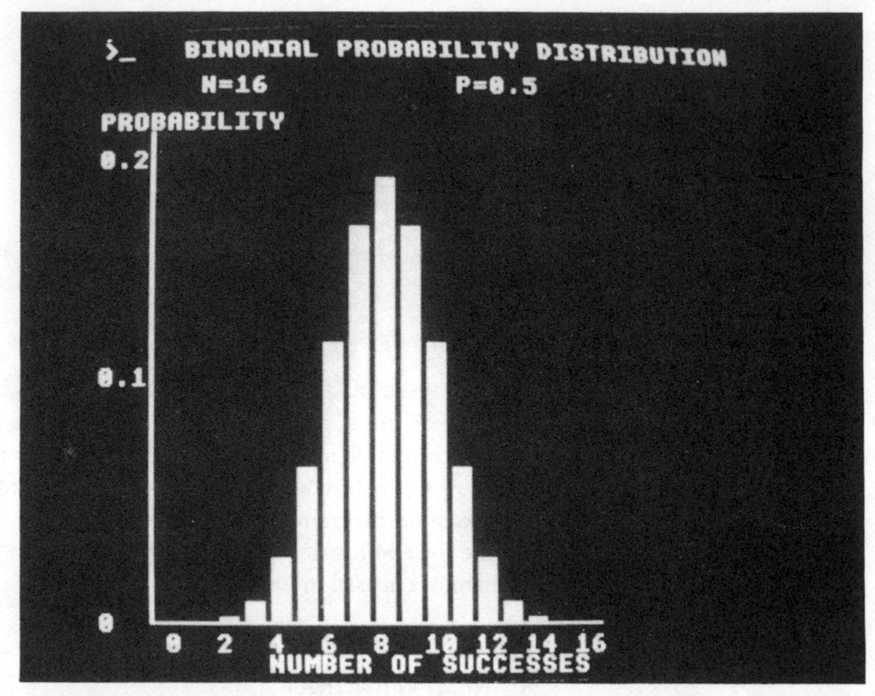

$n = 16$ and $p = 0.5$

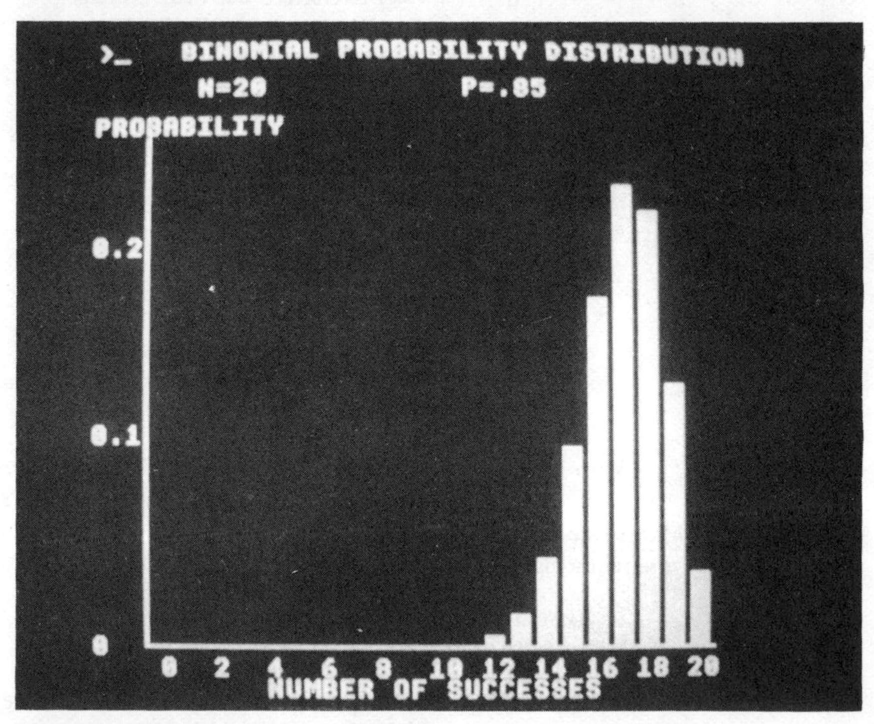

$n = 20$ and $p = 0.85$

Fig. 8.3 (continued next page)

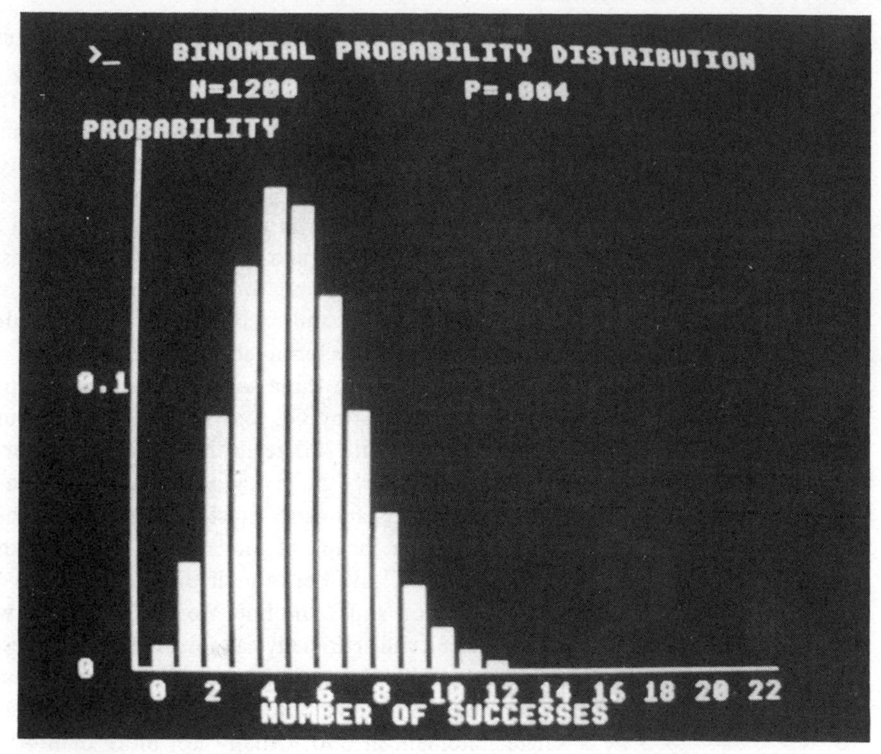

$n = 1200$ and $p = 0.004$

Fig. 8.3 (continued)

A problem can arise if a very large n is used with a p that is not small enough when q^n can be so small that it is out of the range of numbers available on the computer. The first term may thus be computed as zero and thus all terms are given as zero. Even so, values of n of over a thousand can be successfully used with a suitable p. Figure 8.3 shows examples of the display that can be obtained when the output is used with a routine to draw a histogram. We have found such a program invaluable to demonstrate the approach to the normal distribution even when using a microcomputer having no high resolution graphics. With modern machines the graphics facility also allows for the superimposition of a normal curve using different colours if available.

8.7 EXPLORING THE EFFECT OF CLASS SIZE WHEN GRAPHING A SET OF DATA

When a sample of measurements is available, taken from a continuous distribution, the data is often sorted into classes and a histogram drawn to reveal visually the shape of the distribution. The choice of class width has a not inconsiderable influence on the ease with which a pattern may be seen and can also affect the judgement that is made about the quality of the distribution. To sort the same data serveral times in exploring this effect is a time-consuming and tedious process that is not likely to be attempted by hand, especially if the number of data items is at all large. It is natural to use a computer for this task. Since the greatest benefit is obtained by running the program several times using the same data, it is best to use a data file or include the data in the program, it then has to be typed only once. In the example program we give here (Program 8.7) the data is included in a data section at the end that starts with n, the number of items followed by the beginning and end of the range. These values are first read into stores N, XL and XH and then the choice of the number of classes is offered. The class width is H and each data item is read into X and then the class XC to which it belongs is determined in line 220.

The program thus falls into three sections. The first, from line 100, sets the constants and prepares the array CL that will be used to count the number in each class. The second, from line 200, reads the data and finds the class frequencies. The third, from line 300, draws the histogram. This last section will vary between computers and the precise statements must suit the machine being used. We will assume that there is some means of moving the cursor to the bottom of the column to be printed and give here the artificial instruction 'move, c, l' where c is the character position and 1 the line. We also need some way of moving the cursor up the screen and will artificially call this 'cursor up'. These two instructions will need to be replaced by the appropriate statements. In many cases the use of the POKE statement can achieve both these and may replace lines 320, 330 and 335 by a single statement at 330. Usually not more than 40 characters can be printed on one line of the VDU and this puts a limit on the number of columns that can be displayed. With a graphics facility this limitation is removed and the column width may easily be varied to suit the number chosen.

Program 8.7 – To sort data and draw a histogram

```
 90 REM**PROGRAM 8.7**
100 REM***FREQUENCY HISTOGRAM***
110 PRINT "HOW MANY CLASSES";
115 INPUT C
120 READ N,XL,XH
125 H = (XH - XL)/C
130 DIM CL(C)
135 FOR J = 1 TO C
140 CL(J) = 0
145 NEXT J
200 REM***INPUT DATA/COUNT CLASSES***
210 FOR J = 1 TO N
215 READ X
220 XC = INT((X - XL)/H) + 1
225 CL(XC) = CL(XC) + 1
230 NEXT J
300 REM***DRAW HISTOGRAM***
310 FOR K = 1 TO C
315 IF CL(K) = 0 THEN GOTO 345
```

```
 320 Move,K,20
 325 FOR J = 1 TO CL(K)
 330 PRINT "■";
 335 Cursor up
 340 NEXT J
 345 NEXT K
1000 REM***DATA SECTION***
1005 DATA 200,10,35
1010 DATA 14,17,29,22,.......
1020 DATA ........
..............
```

For the BBC computer the program may be completed by the lines:

```
305 CLS
320 P = 20
330 PRINT TAB(2*K,P) CHR$(255)
335 P = P - 1
```

In this example program the statements 'Move' and 'Cursor up' in lines 320 and 335 are artificial and should be replaced by the statements appropriate to the computer being used. Data line 1005 shows that there are 200 data items lying between 10 and 35 and these must be put in the following lines.

9

Randomness, probability and simulation

9.1 WHAT IS A RANDOM NUMBER?

There are many familiar ways of picking random numbers such as rolling dice, picking phone numbers from a telephone directory, putting numbers on balls or pieces of paper and picking them out of a bag – there are even tables of random numbers and most computers have functions that produce random numbers. We are familiar with the mechanisms used, say, to make the draw for the World Cup football competition's final rounds, to select the prize-winners in a raffle and even to choose the winners in national lotteries such as Premium Bonds. Essentially some physical method is used to try to ensure that each of the participants has an equal chance of being selected. Such mechanisms can be applied 'without replacement' in which case once a number has been drawn it cannot subsequently be redrawn (as in the above examples) or 'with replacement' in which case it is returned to the mechanism for possible re-selection at the next draw. The key issue here is thc notaion of 'equal chance'.

For the purposes of this chapter we are concerned with mechanisms for selecting elements of a set of n different objects with equal probability – or, more technically, with sampling from a uniform distribution. Suppose, for example, we want to be able to select a random integer between 0 and 99 inclusive. Some physical means to achieve this are:

(i) The 'top-hat method' – take 100 identical pieces of paper and write each of the numbers 0, 1, 2, . . . , 99 on a different piece. Fold them tightly, put them into a container, such as a top-hat, waste-paper bin etc., shake well and pick out a piece of paper. Note the number written on it.

(ii) Spinners – carefully cut a regular 10-sided polygon from card. Write each of the numbers 0–9 one against each of the sides of the polygon. Pierce the centre of the polygon with a knitting-needle and push a pencil into the hole. If you spin the polygon on the pencil tip it will eventually come to rest with one edge against the horizontal surface. Note the digit written against this edge and record it as, say, the units digit of a 2-digit number. Repeat the operation but this time use the resulting digit as the tens digit of the number.

(iii) Dice and coin – the spinner could be replaced by a die and a coin, but that would give 12 possible outcomes. Suppose we treat the 5 òn the die as a zero and decide that if a 6 is rolled then the die will continue to be rolled again until something other than a 6 appears. We now have a means of selecting the digits, 0, 1, 2, 3, 4 with equal probability (given a fair die). Now toss a coin – if it lands showing a head then treat the digit from the die as it stands but if it shows tails then add 5 to the digit from the die. We now have a means of selecting a digit between 0 and 9 with equal probability and so we can select 2 such digits to make our 2-digit random number.

(iv) Special dice – some manufacturers of educational equipment sell 20-sided dice (icosahedra) with each of the digits 0–9 engraved on two opposite faces, or 10-sided dice (usually truncated octahedra which look 'nearly' regular). Again two throws are needed to generate the 2-digit number.

These methods are all random in the sense that each outcome is quite independent of the previous outcome. There are some methods, though, in which this is not the case but in which the outcome seem to conform to no particular pattern. Such numbers are known as 'pseudo random numbers' and we shall look at a few of the ways of producing such numbers.

9.1.1 Mid-square technique

Take a starting 2-digit number, for example, 76 and call this x_0. Square it to obtain a 4-digit number $x_0^2 = 5776$ and take the middle two digits to obtain the next 'random number' $x_1 = 77$. Square this: $x_1^2 = 5929$ and thus $x_2 = 92$ and so on. This produces a pseudo-random sequence 76, 77, 92, 46, 11, 12, 14, 19, 36, 29 . . . . Unfortunately great care has to be taken to find a suitable starting value. For instance if $x_0 = 75$ then

$$x_0^2 = 5625 \qquad x_1 = 62$$
$$x_1^2 = 3844 \qquad x_2 = 84$$
$$x_2^2 = 7056 \qquad x_3 = 05$$
$$x_3^2 = 0025 \qquad x_4 = 02$$
$$x_4^2 = 0004 \qquad x_5 = 00$$

which is where the sequence sticks.

9.1.2 Mid-product method

Take two initial 2-digit numbers, for example, $x_1 = 34$ and $x_2 = 61$, form their product $p = 2074$ and take x_3 as the middle two digits of p, that is, $x_3 = 07$. Now use $p = x_2 x_3$ to form x_4 and so on.

$$p = x_2 x_3 = 61 \times 07 = 0427 \qquad x_4 = 42$$
$$p = x_3 x_4 = 07 \times 42 = 0294 \qquad x_5 = 29$$
$$p = x_4 x_5 = 42 \times 29 = 1218 \qquad x_6 = 21$$

9.1.3 Lehmer-congruence method

Choose a pair of numbers k and m and form a sequence in which $x_{n+1} = kx_n \pmod m$. Suppose $k = 17$ and $m = 100$ then we could start a sequence with, say, $x_1 = 26$. Then

$$kx_1 = 17 \times 26 = 442 \qquad x_2 = 42$$
$$kx_2 = 17 \times 42 = 714 \qquad x_3 = 14$$
$$kx_3 = 17 \times 14 = 238 \qquad x_4 = 38$$
$$kx_4 = 17 \times 38 = 646 \qquad x_5 = 46 \qquad \text{and so on.}$$

In this example obviously all values in a sequence starting with an odd number will be odd numbers and from an even number will be even numbers. For computer use the technique is applied to numbers with more than 2 digits and much research has gone into finding efficient values for k and m. In BASIC numbers are usally stored in binary floating point form to an accuracy of around 32 binary digits (around 9 decimal digits) and Tocher quotes $k = 7^{11} = 366\,714\,004$ and $m = 2^{29} + 1 = 536870913$ as particularly good choices for this kind of representation.

9.2 RANDOM NUMBERS IN BASIC

Most versions of BASIC include a function RND(X) which calculates the next random number in a sequence of pseudo-random generated by Lehmer congruences. To start such a sequence an initial value x_1 is needed before the recurrence $x_{n+1} = kx_n \pmod m$ can be used. Such an initial value is called the *seed* of the random sequence. Obviously sequences starting with the same seed will be identical and, strangely as it may appear, this property of being able to generate *repeatable* sequences of pseudo-random numbers can be very useful in performing sampling experiments. Usually the values of members x_n of the sequence are divided by the value of m to give numbers in the range $[0, 1)$. There are some variations, though, in the way in which different versions of BASIC use the RND function and in how the process is given its initial seed. Try this simple program (Program 9.1) to see what your microcomputer produces:

Program 9.1 – To test the RND function

```
100  REM PROGRAM 9.1 – TO TEST RND(X)
110  FOR I = 1 TO 20
120     PRINT RND(1)
130  NEXT I
```

Run the program several times to see if it repeats the same set of values. In some versions of BASIC there is a RANDOMIZE command to re-seed the sequence – in this case the same set of random numbers will be produced unless the RANDOMIZE command is inserted before the loop, at line 105, say.

Some versions of BASIC allow you to enter your own seed by using RND with a negative value. In this case try adding these lines to Program 9.1:

```
102  INPUT X
104  Y = RND(X)
```

and run the program with different values of the input X (including some negative ones). Other versions of BASIC allow integer variables as well as floating point variables and in some cases they have adaptations of RND to produce an integer in a given range. For the purpose of this chapter, though, we will assume that RND(1) will return a value in the range $0 \leqslant \text{RND}(1) < 1$.

9.3 SIMULATING DICE

The problem is to map any value of RND(1) from the range $[0, 1)$ onto one of the integers 1, 2, 3, 4, 5 or 6. Fortunately there is a simple and general technique for this process that we call 'stretch, chop and shift'. If you multiply RND(1) by 6 you *stretch* the interval to $[0, 6)$. If then you *chop* by taking the integer part INT(6 * RND(1)) then the result takes one of the values 0, 1, 2, 3, 4, or 5. Finally if you *shift* by adding 1 to the value then 1 + INT(6 * RND(1)) is a member of the required set. To test this try amending line 120 of Program 9.1 to 120 PRINT 1 + INT(6 * RND(1)). We are now in a position to produce a program to investigate totals, say, on rolls of two dice.

Program 9.2 – The totals on two dice

```
100  REM PROGRAM 9.2 – TOTALS OF TWO DICE
110  FOR I = 1 TO 20
120      PRINT 2 + INT(6*RND(1)) + INT(6*RND(1))
130  NEXT I
```

9.4 MONTE CARLO METHODS

These methods involve the use of randomly generated numbers to tackle problems that cannot be adequately approached by the use of the usual analytic methods. A simple application of the approach is the evaluation of definite integrals by approximating areas or volumes. A rectangle of known size is chosen to surround the unknown area. Points are chosen at random within the rectangle by generating random numbers to be used as co-ordinates (imagine throwing darts at a board) and are tested to see if they lie within the unknown area. A count is kept of the number of points generated and of the number of points that fell within the area. The ratio of these two counts gives the fraction of the area of the rectangle occupied by the unknown area and hence that area can be estimated.

As an example consider estimating π by enclosing a circle in a square. We need only consider one quadrant of the circle. In Fig. 9.1 we can easily generate points such as P_1 and P_2 whose co-ordinates satisfy $0 \leqslant x, y < 1$ just by using two values of RND(1) and we can test whether or not such a point lies within the circle by comparing $x^2 + y^2$ with 1. If $x^2 + y^2 < 1$ then P lies within the circle. Suppose we generate N points and S of them lie within the quadrant then

$$\frac{\text{area of quadrant}}{\text{area of square}} = \frac{\pi/4}{1} \approx \frac{S}{N} \quad \text{and so} \quad \pi \approx \frac{4S}{N}$$

Program 9.3 – A Monte Carlo estimate for π.

```
100 REM PROGRAM 9.3 – TO ESTIMATE PI
110 S=0
120 FOR I=1 TO 1000
130    X=RND(1)
140    Y=RND(1)
150    IF X*X+Y*Y<1 THEN S=S+1
160 NEXT I
170 P=4*S/1000
180 PRINT "ESTIMATE FOR PI IS "; P
```

Another well-documented method for estimating π by random methods is Buffon's needle experiment and this can also be easily simulated by a short computer program.

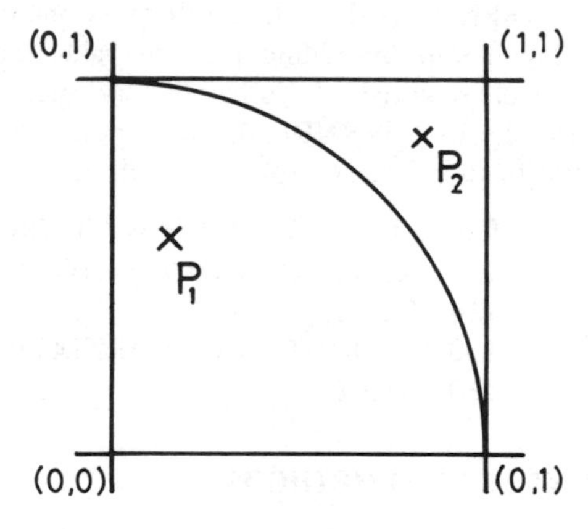

Fig. 9.1

9.5 MARKOV CHAINS

Matrices are often introduced to pupils in connection with networks. Thus the network illustrated might be described by the simple matrix:

$$\begin{array}{cc} & \text{To} \\ & \begin{array}{ccc} A & B & C \end{array} \\ \text{From} \begin{array}{c} A \\ B \\ C \end{array} & \begin{pmatrix} 0 & 1 & 1 \\ 1 & 0 & 1 \\ 1 & 1 & 0 \end{pmatrix} \end{array}$$

B

A C

where each element represents the number of arcs joining any two nodes. Suppose, then, that each node represents some state of a system and that each arc represents a possible transition from one state to another. Then each arc could have a probability associated with it. If these probabilities are constants which are independent of the state of the system then we have what is called a Markov chain.

A teacher has three favourite examination questions A, B and C. He always uses just one of these in the end of the year examination for his class but he never uses the same question two years in a row. If one year he uses question A there is a 50% chance that next year he will use question B, if he uses question B there is a 70% chance that he will use question C the next year and if he uses question C there is a 25% chance that he will use question B the following year. In the long run what are the relative proportions of questions A, B and C that he uses?

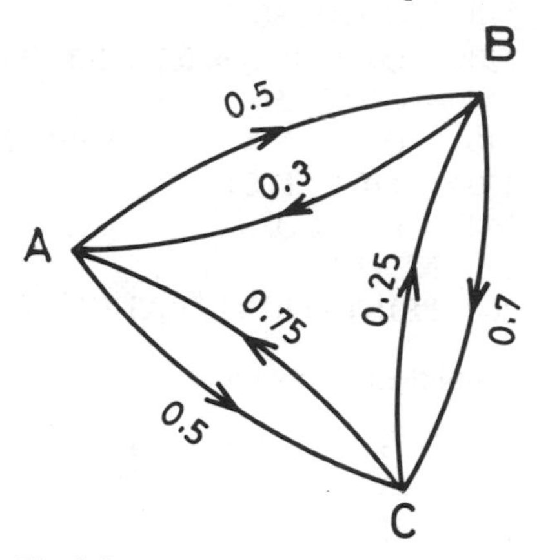

Fig. 9.2

A diagram for this problem is shown in Fig. 9.2 and these probabilities could be represented in a matrix. Such a matrix is called a 'stochastic matrix' and there is not common agreement among books as to which convention to use in labelling rows and columns. We have found it more convenient to use the following convention:

$$\begin{array}{ccc} & & \text{From} \\ & & \begin{array}{ccc} A & B & C \end{array} \\ \text{To} & \begin{array}{c} A \\ B \\ C \end{array} & \begin{pmatrix} 0 & 0.3 & 0.75 \\ 0.5 & 0 & 0.25 \\ 0.5 & 0.7 & 0 \end{pmatrix} \end{array}$$

From this matrix we can see that the probability of a transition from state B to state A is 0.3 and from state B to state C is 0.7. To simulate this problem we shall use the numbers 1, 2, 3 rather than the letters A, B, C. Suppose Q holds the current state of the system then we can start exploring a chain by specifying a starting value for Q. If we choose, say, $Q = 1$ as our initial value then this is equivalent to starting with question A. By now choosing a random number $X = \text{RND}(1)$ between 0 and 1 we could move to state $Q = 2$ if $X \geqslant 0.5$ and to state $Q = 3$ if $X < 0.5$. The next program (Program 9.4) employs this technique.

Program 9.4 – A simple Markov chain

```
100 REM PROGRAM 9.4 – A MARKOV CHAIN
110 DIM T(3)
120 FOR I = 1 TO 3
130    T(I) = 0
140 NEXT I
150 Q = 1 : T(Q) = 1 : PRINT Q,
160 FOR I = 1 TO 99
170    X = RND(1)
180    IF Q = 2 THEN GOTO 220
190    IF Q = 3 THEN GOTO 240
200    Q = 2 : IF X < 0.5 THEN Q = 3
210    GOTO 250
220    Q = 1 : IF X < 0.7 THEN Q = 3
230    GOTO 250
240    Q = 1 : IF X < 0.25 THEN Q = 2
250    T(Q) = T(Q) + 1
260    PRINT Q,
270 NEXT I
280 PRINT : PRINT "RATIOS", T(1);":";T(2);":";T(3)
```

A sample output might start with:

```
1    3    1    3
1    3    2    3
1    2    3    1
3    1    2    3
```

and produce a final output of:

```
RATIOS   37 : 26 : 37
```

There is, of course, an analytic solution to this class of problem which involves finding a 'steady state' vector for the probability matrix (that is, an eigenvector corresponding the eigenvalue of 1) but this kind of simulation can be used for giving experience of the behaviour of this kind of problem which may be a useful precursor to developing a theoretical approach.

9.6 SAMPLING FROM A DISTRIBUTION

Many queueing problems are suitable for investigation by computer simulation The following example is taken from the section on 'Queueing and Simulation' By Dr S. H. Hollingdale in the Penguin book *Newer Uses of Mathematics* and concerns a medical appointments system. A doctor wishes to start an appointments system for his two-hour surgery but does not know what is the best gap to leave between successive appointments. In order to investigate this it is necessary to collect data about the variability of the time the doctor spends with each patient – the consultation time. This data would have to be collected over some time to even out many particular irregularities. Suppose this has been done for a month, involving some 800 consultations and that the results are shown in Fig. 9.3.

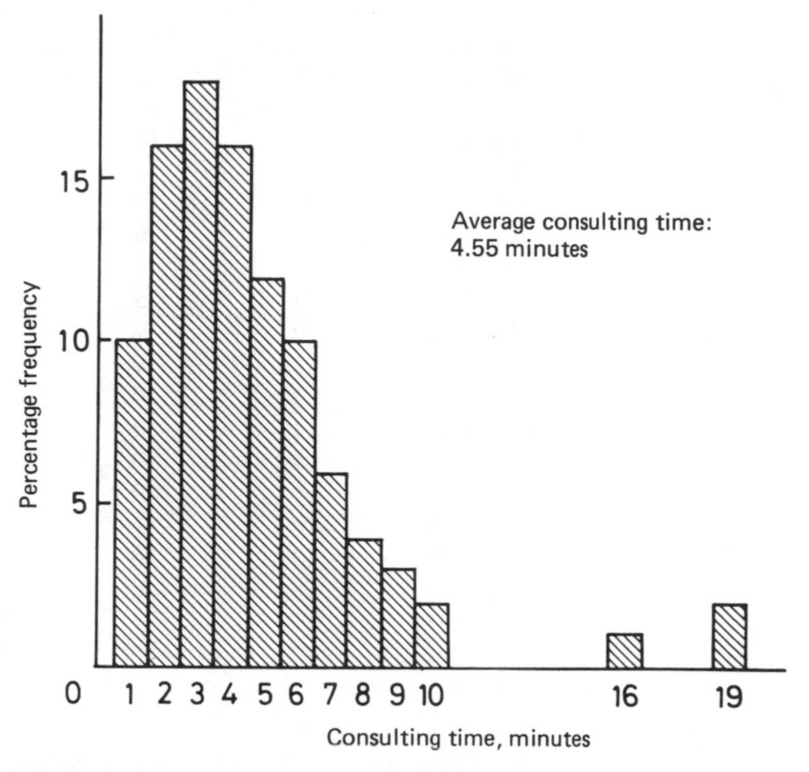

Fig. 9.3

From this histogram we can see that the modal consultation time is 3 min, from calculations we find that the *mean* time is 4.55 min and, by considering cumulative frequencies, that the *median* time is 4 min. The problem, then, is to determine a suitable interval between appointments so that the doctor is occupied as fully as possible but that unacceptable delays do not build up in the waiting room. Suppose we investigate a surgery starting at 10 a.m., finishing at 12 noon, and in which only patients who have been given appointments arrive at the surgery and that they do so on time and without fail! (Further complications can easily be built into the model at a later stage.) If we choose 5 min as a suitable gap between appointments then patients will arrive at 10.00, 10.05, 10.10, ... until the last appointment arrives at 11.55. When a simulated patient arrives at the surgery we need to take a random sample from the observed distribution of consultation times to decide how long he or she spends with the doctor. To do this we use the histogram of Fig. 9.3 to draw up a frequency distribution table as shown in Table 9.1.

Table 9.1

Time interval (min)	Mean time (min)	% frequency of consulting time	Cumulative % frequency	Random number range
½–1½	1	10	10	00–09
1½–2½	2	16	26	10–25
2½–3½	3	18	44	26–43
3½–4½	4	16	60	44–59
4½–5½	5	12	72	60–71
5½–6½	6	10	82	72–81
6½–7½	7	6	88	82–87
7½–8½	8	4	92	88–91
8½–9½	9	3	95	92–94
9½–10½	10	2	97	95–96
10½–11½	11	0	97	
11½–12½	12	0	97	
12½–13½	13	0	97	
13½–14½	14	0	97	
14½–15½	15	0	97	
15½–16½	16	1	98	97
16½–17½	17	0	98	
17½–18½	18	0	98	
18½–19½	19	2	100	98–99
19½–20½	20	0	100	

In order to generate a random sample from this pattern of consultation times we use INT(100*RND(1)) to generate a random two-digit number in the range 00–99. From the final column of Table 9.1 we can use such a number to map onto a mean consultation time in column 2. Thus the random number 37 lies between 26 and 43 inclusive and so corresponds to a consultation time of 3 min. This is the computer equivalent of a top-hat method in which 100 pieces of paper are put into the hat – 10 pieces of paper have a 1 written on them, 16 have a 2, 18 have a 3 and so on. Thus although each piece of paper has an equal chance of being picked we have weighted the probabilities of the outcomes in the ratio of their frequencies. This technique is widely used to sample from empirically derived distributions. A subroutine can now be produced to sample from Table 9.1.

```
1000  REM SUBROUTINE 9.5A – TO FIND A CONSULTING
      TIME T
1010  X = INT(100*RND(1))
1020  IF X < 10 THEN T = 1 : RETURN
1030  IF X < 26 THEN T = 2 : RETURN
1040  IF X < 43 THEN T = 3 : RETURN
```

```
1050  IF X < 60 THEN T = 4 : RETURN
1060  IF X < 72 THEN T = 5 : RETURN
1070  IF X < 82 THEN T = 6 : RETURN
1080  IF X < 88 THEN T = 7 : RETURN
1090  IF X < 92 THEN T = 8 : RETURN
1100  IF X < 95 THEN T = 9 : RETURN
1110  IF X < 97 THEN T = 10 : RETURN
1120  IF X < 98 THEN T = 16 : RETURN
1130  T = 19 : RETURN
```

A sample run of one surgery can now be performed. First, though, we need to decide what statistics to collect about the simulated surgery to be used in comparison with other simulations using different appointment gaps. We are interested in the number of patients seen, the distribution of their waiting times, the amount of time during the surgery that the doctor is idle and the finishing time of the surgery. We might also want to compare the sampled consultation times with the original distribution to see how typical a surgery we have simulated. In the following program we use 4 arrays. $C(I)$ holds the sampled consultation time for the Ith patient, $E(I)$ holds the time since the surgery started that he enters the doctor's consulting room, $D(I)$ the time in minutes after 10 a.m. that he departs and $W(I)$ the number of minutes he spends waiting. G is the gap between appointments, U is the number of minutes that the doctor spends unoccupied, N is the number of patients seen and TW is a running total of waiting times. To gain a better picture, though, we need to simulate many surgeries. So once the program has been run a few times to check that is produces sensible output then lines 160, 170 and 300 can be deleted and the body of the program put in a loop:

```
135  FOR J = 1 TO 20
335  NEXT J
```

Program 9.5 – To simulate a doctor's surgery

```
100  REM PROGRAM 9.5 – TO SIMULATE A SURGERY
110  G = 5
120  N = INT(120/G)
130  DIM C(N), D(N), W(N), E(N)
140  U = 0 : GOSUB 1000
150  C(1) = T : D(1) = T : E(1) = 0 : W(1) = 0 : TW = 0
160  PRINT "PATIENT"; TAB(10); "ARRIVE"; TAB(17);
     "ENTER"; TAB(24); "DEPART"; TAB(32); "WAIT"
170  PRINT : PRINT 1; TAB(10); 0; TAB(17); 0; TAB(24); T;
     TAB(32); 0
180  FOR I = 2 TO N
190     A = G*(I - 1) : REM Arrival time
200     GOSUB 1000 : C(I) = T : REM Sample consulting time
210     IF A > D(I - 1) THEN GOTO 260 : REM No-one being seen
220     E(I) = D(I - 1) : REM Entry time = previous departure
```

```
230     W(I) = E(I) – A : REM Waiting time = entry – arrival
240     TW = TW + W(I)
250     GOTO 290
260     E(I) = A : REM Entry time = arrival
270     W(I) = 0
280     U = U + A - D(I - 1) : REM Doctor may have been waiting
290     D(I) = E(I) + C(I)
300     PRINT I; TAB(10); A; TAB(17); E(I); TAB(24); D(I);
        TAB(32); W(I)
310   NEXT I
320   PRINT : PRINT N; "PATIENTS", U; "MINS. UNOCCUPIED"
330   PRINT TW/N; "AVERAGE WAIT", D(N)-120; "MINS.
      OVERRUN"
340   END
```

Thus we can simulate 20 surgeries with an appointment gap of 5 min. By amending line 110 to G = 6 or G = 4 we can study the effects of lengthening or shortening the gap between arrivals. Of course other statistics could easily be selected and the model can be made more sophisticated by including distributions for lateness of arrival, probability of non-arrival, probability of emergency additions etc.

This kind of simulation is one of the main techniques in the body of applications of mathematics usually referred to as OR or Operational Research and has been used as the basis for many major planning decisions.

9.7 SAMPLING FROM STANDARD DISTRIBUTIONS

The pseudo-random numbers generated by the RND function should be uniformly distributed in the range $[0, 1)$ but frequently we may wish to generate random samples from other common distributions. We now give (mainly without derivation) some algorithms for generating such samples – further details may be found in Knuth (Vol. 2) [14] or Tocher [20].

9.7.1 Standard normal distribution

The central-limit theorem tells us that the mean of n random samples from the distribution will, as n tends to infinity, approach a normal distribution. So by adding n samples from the uniform distribution $U(0, 1)$ the resulting values are approximately normally distributed and have mean $n/2$ and standard deviation $\sqrt{n/12}$. Thus adding 12 uniform variates together and subtracting 6 yields a value which is distributed approximately normally with mean 0 and standard deviation 1. This is an example in which 12 is quite a reasonable approximation to infinity!

Program 9.6 – A subroutine to sample from a normal distribution

```
1000  REM PROGRAM 9.6 – SUBROUTINE FOR NORMAL
      DISTRIBUTION 1
1010  V = -6
1020  FOR I = 1 TO 12
```

```
1030     V = V + RND(1)
1040  NEXT I
1050  RETURN
```

Of course the value of V will lie in the range [-6,6) and so this method is not suitable if the tails of the distribution are critical.

For better approximations we need to find the inverse function for approximations to the cumulative distribution function of the normal distribution. Suppose U is a sample from the uniform distribution $U(0,1)$ then:

$$x = \pm\sqrt{\frac{\pi}{8}\ln\frac{1+U}{1-U}}$$

is a good approximation to a normal variate. In implementing this we need to attach a random + or − sign.

Program 9.7 – An alternative method for obtaining normally distributed samples

```
1000  REM PROGRAM 9.7 – SUBROUTINE FOR NORMAL
      DISTRIBUTION 2
1010  K = SQR(3.14159/8) : REM THIS SHOULD BE COMPUTED
      AND STORED IN THE MAIN PROGRAM
1020  U = RND(1)
1030  V = K * LOG((1 + U)/(1 - U))
1040  U = RND(1)
1050  IF U < 0.5 THEN V = -V
1060  RETURN
```

9.7.2 Negative exponential distribution

The probability density function (p.d.f.) of the negative exponential distribution with mean 1 is

$$p(x) = e^{-x}$$

Thus its cumulative distribution function (c.d.f.) is given by

$$P(x) = \int_0^x p(t)\,dt = \int_0^x e^{-t}dt = \Big[-e^{-t}\Big]_0^x = 1 - e^{-x}\ .$$

Now to invert this function put $y = 1 - e^{-x}$ then we have

$$e^{-x} = 1 - y \qquad -x = \ln(1-y) \qquad x = -\ln(1-y)$$

Hence the variate $x = -\ln(1 - U)$ is negative exponentially distributed where U is a uniform variate. However, if U is uniformly distributed so is $1 - U$ and thus $x = -\ln(U)$ will do just as well. If the distribution has mean M then the required function is $x = -M.\ln(U)$

Program 9.8 – To produce samples from a negative exponential distribution

```
1000 REM PROGRAM 9.8 – SUBROUTINE FOR NEGATIVE
     EXPONENTIAL WITH MEAN M
1010 X = RND(1)
1020 V = –M * LOG(X)
1030 RETURN
```

Note. For most microcomputers the function LOG(X) generates natural logarithms $\ln x$, but on a few it generates logarithms to the base 10 and the required function is LN(X).

9.7.3 Poisson distribution

This is a discrete distribution with the form $P(x = n) = (h^n/n!)\,e^{-h}$ which has both mean and standard devivation equal to h. The technique involves generating independent samples from a negative exponential distribution of mean $1/h$ and adding them together until their total just exceeds h. If that requires the addition of N such samples then $N-1$ is a Poisson sample.

Program 9.9 – Generating Poisson samples

```
1000 REM PROGRAM 9.9 – SUBROUTINE FOR POISSON
     VARIATE – MEAN H
1010 N = 0 : M = 1/H : S = 0
1020 S = S - M * LOG(RND(1))
1030 N = N + 1
1040 IF S <= 1 THEN GOTO 1020
1050 V = N - 1
1060 RETURN
```

9.7.4 Chi-squared distribution

A sample value of χ^2 on n degrees of freedom is found by summing the squares of n standard normal variates.

Program 9.10 – Chi-squared samples

```
1000 REM PROGRAM 9.10 – SUBROUTINE FOR
     CHI-SQUARED WITH N D. F.
1010 V = 0 : K = 3.14159/8
1020 FOR I = 1 TO N
1030   U = RND(1)
1040   T = LOG((1 + U)/(1 - U))
1050   V = V + K * T * T
1060 NEXT I
1070 RETURN
```

9.8 USING DISTRIBUTIONS

The subroutines for sampling from probability distributions can be applied to many simulations of random events such as may be found in biology, chemistry and physics. As an example of their use in mathematics we can investigate the distribution of the means of sample of a given size from a particular distribution. Suppose we draw groups of N samples from a standard normal distribution and investigate the distribution of their means M.

Program 9.11 – To investigate standard error

```
 100 REM PROGRAM 9.11 – STANDARD ERROR
 110 K = SQR(3.14159/8)
 120 INPUT "GROUP SIZE", N
 130 INPUT "NO. OF SAMPLES", P
 140 X1 = 0 : X2 = 0
 150 FOR I = 1 TO P
 160   S = 0
 170   FOR J = 1 TO N
 180     GOSUB 1000
 190     S = S + V
 200   NEXT J
 210   M = S/N : X1 = X1 + M : X2 = X2 + M*M
 220 NEXT I
 225 M = X1/P : D = SQR(X2/P - M*M)
 230 PRINT "MEAN="; M, "S.D.="; D
 240 END
1000 REM***NORMAL SAMPLE****
1010 U = RND(1)
1020 V = K*LOG((1 + U)/(1 - U))
1030 U = RND(1)
1040 IF U<0.5 THEN V = -V
1050 RETURN
```

Note that on a BBC micromputer line 1020 would need LN in place of LOG to give natural logarithms. Using this program, which employs the techniques introduced in Programs 8.1, 8.2 and 9.7, we can perform several runs and, with luck, see that the results conform to the theoretical result – that the mean value M approximates the mean, 0, of the parent distribution, but that the standard deviation D approximates the standard deviation, 1, of the parent distribution divided by $\sqrt{N}$. Try producing similar programs to investigate other 'standard' results such as the Central Limit Theorem.

10

Internal number representation and computer accuracy

Computers have a reputation for being utterly reliable and completely accurate. It is true that mistakes occur so seldom that it can safely be assumed that they will not be made, and there are various built in checks to detect them in the unlikely event of there being one. However, there is a limit to the size of computer stores and calculations are performed within the limitation that this places on them. It is necessary to have some understanding of the way numbers are represented to be able to avoid errors, or to work within the restrictions that they impose on us.

10.1 NUMBERS IN A COMPUTER

It is well known that in a computer numbers are kept in binary form, whereas we always use base ten. Thus 39 is stored as 100111_2 and 44 as 101100_2. Each binary figure is called a *bit* and clearly many more bits are required than the number of decimal figures. In the computer storage the bits are grouped together with 8 bits making 1 byte and this is the unit of memory that is usually used to denote the memory size. In many microcomputers the storage allocated to one number is 4 bytes (that is, 32 bits) so that the largest exact integer that can be stored is the binary number composed of 32 1's which is 4 294 967 295 in base ten. Longer numbers are corrupted when they are stored and are either chopped or rounded to fit into the available space. This is one source of error in calculation.

There are two types of number representation used in computers: *integer* and *floating point.* Integers are usually identified by a % sign in the store name, thus in a program B% will contain an integer whereas B will contain a floating point number. In some systems only certain variables may be designated as integers. Floating point numbers are stored in exponent form, thus there are two parts to the number; a binary decimal starting 0.1 . . . and the power of two by which this must be multiplied to give the true value. Thus our example of 39 becomes 0.100111_2 with exponent 6 or rather 110_2 since this too will be in binary form. The limitation on store size means that there is some upper limit to the exponent. On some microcomputers this is 127 so that the largest floating point number that the computer can cope with is

$$0.\underbrace{1111\ldots1}_{32\ 1\text{'s}}{}_2 \times 2^{127}$$

and in base ten this is

$$1.701\,411\,83 \times 10^{38}$$

Any attempt to use a number larger than this will produce an error message. In a similar way there is a limit to how small a number can be and in some cases is

$$0.1 \times 2^{-129}$$

or

$$2.938\,735\,88 \times 10^{-39}$$

Any calculated number that is smaller than this will be stored as zero.

10.2 HEXADECIMAL NUMBERS

Because the binary numbers used are so long, and there is no simple relationship between decimal and binary, another number base is often used in communicating with a computer. This is base sixteen and is chosen because it fits the organisation of computer storage into bytes. Half a byte, called a 'nibble', is four bits and can contain any binary number from 0000_2 to 1111_2, that is, 0 to 15. Base sixteen (or hexadecimal) numbers need the digits 0 to 15 to represent them. Since we need a single symbol for each digit we use A, B, C, D, E and F for the numbers 10, 11, 12, 13, 14 and 15. As examples of hexadecimal representation 39 is two sixteens plus seven and so becomes 27_{hex}, similarly 44 becomes $2B_{hex}$. There is a particularly simple relationship between hexadecimal and binary representations. Since $16 = 2^4$ each hexadecimal place corresponds to four binary places and the correspondence between the hexadecimal digits and binary numbers is given in the following table.

Hex digit	0	1	2	3	4	5	6	7
Binary number	0000	0001	0010	0011	0100	0101	0110	0111

Hex digit	8	9	A	B	C	D	E	F
Binary number	1000	1001	1010	1011	1100	1101	1110	1111

It is now clear that each nibble contains one hexadecimal digit, and a byte contains two hexadecimal digits. Any binary number can easily be changed to hexadecimal by replacing each group of four bits by the corresponding hexadecimal digit. Thus

$$39 = 0010\,0111_2 = 27_{hex}$$

and

$$44 = 0010\,1100_2 = 2C_{hex}$$

also

$$101\ 1001\ 0010\ 1011\ 1000_2 = 592B8_{hex}.$$

Sometimes it is useful to enter numbers directly into the computer in hexadecimal form. The way to do this varies from one machine to another, but sometimes hexadecimal numbers are identified by the symbol '&' at the start of the number. A system for programming your own characters that is sometimes available can be simplified by using hexedecimal numbers. In the method to be described here each character is composed of a grid of 64 pixels in eight rows of eight. Each pixel corresponds to one bit in a binary number and will be coloured if that bit is a 1 or black if it is a 0. Each row of the desired character has to be converted into a number and the eight numbers entered into the computer. Decimal or hexadecimal numbers may be used but the conversion from binary to the latter is easier. As an example, suppose we wish to produce a diagonal arrow symbol. In the illustration the corresponding binary numbers are given alongside each row. To convert these to decimal involves a lot of calculation and carries with it the risk of error. If hexadecimal numbers can be used then each block of four bits converts directly to one hexadecimal digit so that the complete character is represented by eight two-digit hexadecimal numbers. These can then be used directly in the appropriate format to define the character.

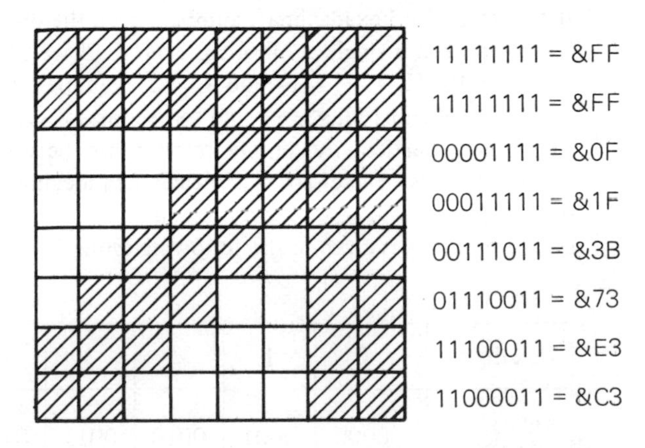

Fig. 10.1 – Programming a character in hexadecimal.

10.3 ERRORS IN CALCULATION

We have seen that a computer rounds numbers to a fixed number of binary digits for storage. This happens to the result at each stage in a calculation, the number first being calculated to more places, then rounded (or chopped) to fit the store. With 32 binary places the result is still so accurate that in most simple calculations the error is of no consequence hence the computer's reputation for high accuracy. However, if a very large number of steps are taken, or the process is particularly sensitive to small changes in the numbers, then serious errors can occur. The control of the build-up of errors is an important aspect of numerical work.

To illustrate the effect in a simple way we imagine a decimal machine that rounds numbers to three figures so that a result that should be 3.9281 becomes 3.93. Consider a few examples of how such a machine would perform.

Example 1. $2.74 \times 58.2 \times 0.876$

First stage: $2.74 \times 58.2 = 159.468$ but the machine now stores this as 159

Second stage: $159 \times 0.876 = 139.284$ stored as 139.

The machine gives a final three-figure answer of 139 though exact calculation yields 139.693 968 that, to three figures, should be 140. With even only two stages in the calculation errors are starting to build up.

Example 2. $7.86 + 4.91 + 5.36$

First stage: $7.86 + 4.91 = 12.77$ rounded to 12.8
Second stage: $12.8 + 5.36 = 18.16$ rounded to 18.2

The final answer is given as 18.2 instead of 18.1 as it ought to be to three figures.

Example 3. $\dfrac{387 \times 2.49}{49.3}$ which should be 19.5 to three figures.

First stage: $387 \times 2.49 = 963.63$ rounded to 964
Second stage: $964 \div 49.3 = 19.553\,753$ rounded to 19.6

However, if this example is organised differently we get:
First stage: $387 \div 49.3 = 7.8499$ rounded to 7.85
Second stage: $7.85 \times 2.49 = 19.5465$ rounded to 19.5

which is correct. A last example demonstrates the disastrous effect of subtracting nearly equal numbers when working to a fixed number of figures.

Example 4. $\dfrac{8.01^2 - 8^2}{0.01}$

The machine first evaluates the squares and stores the results so that the expression becomes $\dfrac{64.2 - 64}{0.01}$ and this is evaluated as 20. The correct answer of 16.01 is quite different. It is interesting to note that in this case rewriting the expression as $\dfrac{(8.01 + 8)(8.01 - 8)}{0.01}$ leads to 16 which is a much better result.

Sometimes it is possible to organise the calculation in a way that controls the errors as this example demonstrates.

The subject of error estimation and control in numerical work is a large and important one that demands a high level of mathematical knowledge. However, even in elementary work, we believe that pupils should become aware of its existence and obtain their own evidence. We give now some examples of error build up in calculations on a computer with 32-bit storage and rounding in calculations. The reader will be able to devise other situations where errors build up to an observable extent.

10.4 NUMERICAL INTEGRATION WITH AN INCREASING NUMBER OF STRIPS

Since $\ln 5 = \int_1^5 \frac{1}{x} dx$ we may seek to use a numerical method to find the area under the graph of $1/x$ as a means of calculating the logarithm. The trapezoidal rule could be used for this, and can easily be understood so is a good choice for elementary students. Increasing the number of strips should show convergence to the required value. The calculation was carried out on a BBC microcomputer with the following result.

Number of strips	Area
50	1.609 949 57
100	1.609 565 89
200	1.609 469 92
500	1.609 443 03
1000	1.609 439 19
2000	1.609 438 21
5000	1.609 438 08
10000	1.609 438 04
20000	1.609 438 14

As an example of convergence to the limit as the number of strips is increased this example performs well at first, but by taking an excessively large number of strips the last value seems to be moving away from the limit. Calculation errors are now beginning to build up and mask the effect being studied. There can be two possible explanations for this. The adding together of 20 000 areas each subject to an error in calculation can cause accumulation of errors in the sum that is significant at this level of accuracy. However, there is another, and perhaps more likely, explanantion. In the method used, each value of x is found by adding the strip width to the previous value. The strip width is calculated as $(5-1)/20\,000$ which is not an exact binary number and is therefore rounded. By the time we get to the final value of x this error will have been included 20 000 times. Random errors tend to cancel out to a certain extent, but in this case, since the same number is used repeatedly, all errors are the same and can accumulate seriously so that the x values used for the final strips are corrupted.

10.5 FINDING A DIFFERENTIAL FROM FIRST PRINCIPLES

Functions such as sine and cosine are evaluated numerically in the computer. These evaluations also involve some errors. The derivative of the function $f(x)$ at $x = a$ has the value which is the limit of $[f(a+h)-f(a)]/h$ as h approaches 0. We may calculate this fraction for a series of successively smaller values of h, but if we attempt to take the process too far we run into the problem of subtracting two nearly equal numbers and the large relative error that results. As an example of this consider the differential of cos x which should be $-\sin x$. If we use $x = \pi/6$ the result should be $-\sin \pi/6$ which is -0.5. Example Program 10.1 was used for this to calculate the fraction $[\cos(x+h) - \cos(x)]/h$ with $x = \pi/6$ and a series of successively smaller values of h. The results are given below the program listing and show that at first reducing h produces convergence towards the expected value. Further reduction in h produces calculation errors that begin not only to swamp the effect being studied, but also to completely spoil the number being calculated. The program may of course easily be modified to study a similar effect for other functions.

Program 10.1 – Numerical calculation of the derivative of $\cos x$ ***at*** $x = \pi/6$. (Note: On some computers the number PI is represented by the symbol π.)

```
100 REM**NUMERICAL DIFFERENTIATION**
101 REM**COS(X) AT X = PI/6**
110 X = PI/6
120 H = 0.1
130 S = (COS(X+H) - COS(X))/H
140 PRINT H, S
150 FOR K = 1 TO 1000 : NEXT K
160 H = H/2
170 GOTO 130
```

Sample output achieved with a BBC microcomputer.

H	S
0.1	−0.542 432 277
0.006 25	−0.502 703 004
2.441×10^{-5}	−0.499 992 371
1.526×10^{-6}	−0.499 877 930
9.537×10^{-8}	−0.498 046 875
5.960×10^{-9}	−0.429 687 500

10.6 SQUARE ROOTS AND SQUARING

Another demonstration of error build up can be made by using the square root repeatedly then squaring the same number of times to try to recover the original number. Program 10.2 may be used for this and provides, in input line 115, for a choice of starting number and the number of times the square root is to be applied. The loop in lines 120 to 130 takes the square root of the starting number C times, then the loop in 150 to 160 squares the result C times and so, without error, would recover the original number. The sample output given was obtained with a BBC microcomputer and shows what can happen. With increasing values of C the errors build up in quite a dramatic way and reach the point where the whole calculation is quite meaningless. Other starting values and other computers can be expected to behave in a similar way.

Program 10.2 – To investigate the effect of square rooting C times, then squaring C times

```
100 REM**SQU.ROOT/SQUARE REPEAT**
110 PRINT "WHAT STARTING NUMBER"
111 PRINT "AND NUMBER OF APPLICATIONS?"
115 INPUT N, C
120 FOR J = 1 TO C
125    N = SQR(N)
130 NEXT J
140 FOR J = 1 TO C
145    N = N*N
150 NEXT J
160 PRINT N
```

Example output starting with $N = 5$ and obtained on a BBC microcomputer.

Value of C	Value of N recovered
5	5.000 000 04
10	4.999 999 84
20	5.002 005 22
30	4.481 621 06
32	7.389 053 64
34	2980.954 01

The corruption that takes place with a large number of applications of the square root illustrates how, in some cases, errors can make a calculation completely invalid. The reader will doubtless be able to find other processes in which errors build up, and that can be used to gain experience of the dangers of assuming that a computer will always be accurate. Experience of this sort also provides a background for the study of errors in numerical analysis, and helps to give some motivation for what is undoubtedly a difficult, but important, area of mathematics.

11

Sequences and limits

The idea of a limit is a fundamental one in mathematics but, owing to the amount of arithmetic involved, there is often little numerical experience at an elementary stage to underpin the later analytical development. By using a computer and some simple programs, a feeling for the topic can be developed with only an elementary background in mathematics as a basis. With the need for tedious calculation removed, the learner can concentrate on the pattern of the resulting numbers.

11.1 RECURRENCE RELATIONS

Consider the sequence of numbers generated by the rule:

$$u_n = \frac{6}{5-u_{n-1}} \qquad (11.1)$$

where each u_n is obtained from the previous term u_{n-1}. If we start with $u_1 = 1$ then $u_2 = 1.5, u_3 = 1.714\,286$ etc. The calculation soon gets tedious so to continue the sequence we can make use of a simple program such as Program 11.1.

Program 11.1 – To generate a sequence by a recurrence relation

```
 90  REM**PROGRAM 11.1**
100  REM**RECURRENCE RELATION**
110  PRINT "WHAT FIRST NUMBER?"
```

```
115  INPUT U
120  N = 1
200  REM**CALCULATION SECTION**
210  PRINT "TERM ";N;"IS";U
220  UP = U
230  U = 6/(5 - UP)
235  N = N + 1
240  FOR J = 1 TO 1000 : NEXT J
250  GOTO 210
```

In the program the current term of the sequence is U and UP is the previous term. At line 220 UP takes the value of the current term ready for a new term to be calculated. Line 240 is to cause the computer to wait before going on to the next term and the value of 1000 may need to be changed to suit the computer speed and the requirements of the particular user. No END is provided and program execution should be stopped from the keyboard. Various starting values for u_1 can be tried and the resulting limit investigated. We can also attempt to find the limiting value another way. If the rule leads to a limiting value L, then applying the rule to L will produce the value L again so that:

$$L = \frac{6}{5-L} \tag{11.2}$$

This equation may be solved and the result related to the performance of the recurrence relation.

Other recurrence relations may be explored by changing line 230 of the program. Another rule, related to the one given above is

$$u_n = \sqrt{5u_{n-1}-6} \tag{11.3}$$

which if it leads to a limit L then L satisfies the equation

$$L = \sqrt{5L-6} \tag{11.4}$$

which leads to the same quadratic equation as (11.2). The different behaviour of the two relations (11.1) and (11.2) can be compared and possibly explained.

Other recurrence relation formulae may be suggested and investigated. Two or more terms may be used to generate the new term. Program 11.1 needs little modification to cope with this. For example if

$$u_n = \frac{3u_{n-1}}{u_{n-2}+2} \tag{11.5}$$

then we need to keep two values at each stage to be able to calculate the next term of the sequence. The lines to be changed or added to Program 11.1 are given as Program 11.2. In line 220 the previous value is put in UP and the one before that in UQ.

Program 11.2 – Lines to be changed or added to Program 11.1 for recurrence relation of equation (11.5)

```
 90 REM**PROGRAM 11.2**
115 INPUT UP
117 PRINT "WHAT SECOND NUMBER?";
118 INPUT U
220 UQ = UP : UP = U
230 U = 3*UP/(UQ + 2)
```

The reader should try several different rules and a variety of starting values for each. In some cases the result depends on the starting values chosen. The following examples are suitable but do not all behave in the same way.

(a) $$u_n = \frac{5u_{n-1}}{u_{n-2} + 2}$$

(b) $$u_n = \frac{u_{n-1} + 2u_{n-2}}{3}$$

(c) $$u_n = \frac{u_{n-1}^2 + 1}{u_{n-2} + 5}$$

(d) $$u_n = \sqrt{u_{n-1}u_{n-2}}$$

11.2 THE YUM-YUM ICE CREAM COMPANY

The Yum-Yum ice cream company assesses the market for its product. There are 1 000 000 customers buying ice cream in its area, and of these 600 000 are currently buying their brand. However, the market is changing and it is estimated that two-fifths of their customers will switch to other brands by next year. On the other hand, they estimate that only one-tenth of those buying other brands will switch to Yum-Yum. Assuming that the market continues in this way we may model the situation by the recurrence formulae

$$\begin{aligned} x_n &= \frac{3}{5}x_{n-1} + \frac{1}{10}y_{n-1} \\ y_n &= \frac{2}{5}x_{n-1} + \frac{9}{10}y_{n-1} \end{aligned} \tag{11.6}$$

where x_n is the number buying Yum-Yum and y_n those buying other brands in year n. Starting with $x_1 = 600\,000$ and $y_1 = 400\,000$ equations (11.6) give

$$x_2 = \frac{3}{5} \times 600\,000 + \frac{1}{10} \times 400\,000 = 400\,000$$

$$y_2 = \frac{2}{5} \times 600\,000 + \frac{9}{10} \times 400\,000 = 600\,000$$

These values may be used to obtain x_3 and y_3 in the same way. Program 11.3 may be used to do the calculations, and to find the expected numbers of customers for several years ahead so that the trend may be explored.

Program 11.3 – To predict the customer numbers for the Yum-Yum ice cream company

```
 90 REM**PROGRAM 11.3**
100 REM**YUM-YUM CUSTOMERS**
110 PRINT "HOW MANY CUSTOMERS IN YEAR 1?"
115 INPUT X
120 Y = 1 000 000 - X
125 N = 1
200 REM**CALCULATION SECTION**
210 PRINT "YEAR ";N;" CUSTOMER NUMBERS"
220 PRINT "YUM-YUM:–"; X; "OTHER:–"; Y
230 XP = X : YP = Y
235 X = 0.6*XP + 0.1*YP
240 Y = 0.4*XP + 0.9*YP
245 N = N + 1
250 FOR J = 1 TO 1000 : NEXT J
255 GOTO 210
```

Those used to using matrices will have noticed that equations (11.6) can be expressed as

$$\begin{bmatrix} x_n \\ \\ y_n \end{bmatrix} = \begin{bmatrix} \frac{3}{5} & \frac{1}{10} \\ \\ \frac{2}{5} & \frac{9}{10} \end{bmatrix} \begin{bmatrix} x_{n-1} \\ \\ y_{n-1} \end{bmatrix} \tag{11.7}$$

in which the numbers in the columns of the matrix add up to 1. The long-term trend that emerges from the calculation is connected with the eigenvalues and eigenvectors of the matrix. Other similar situations may be explored in this way including those with three or more components in the vector.

11.3 THE LIMIT OF A FUNCTION

Functions such as $\dfrac{x^2-3x}{x^2-9}$, $\dfrac{\sqrt{x^2+7}-4}{x-3}$ and $\dfrac{\sqrt{x^2+7}-4}{x^2-6x+9}$ have at least one value of x for which the expression is indeterminate. In the case of these examples for $x = 3$ each becomes 0/0 which is meaningless. The question then arises as to what happens as x gets closer and closer to 3. The exploration is helped by using a program in which the function is evaluated for x equal to $3+h$ with h taking successively smaller and smaller values. The resulting sequence indicates the behaviour of the function as x approaches 3. Program 11.4 evaluates the function for a succession of values of X closer and closer to the chosen value XL of the x-limit. H is the difference of X from XL and at each step the value of H is halved.

The function is defined in line 110 and the example uses $(x^2 - 3x)/(x^2 - 9)$ which is not defined for $x = 3$. When the progam is run, 3 can be input for the x-limit and 1 chosen as the initial value of H. The resulting output is:

```
X = 4          F(X) = 0.571 429
X = 3.5        F(X) = 0.538 462
X = 3.25       F(X) = 0.520 000
X = 3.125      F(X) = 0.510 204
X = 3.0625     F(X) = 0.505 155
```

and just a few more terms provide evidence of convergence to 0.5. By using a negative value for H the behaviour as x approaches 3 from below can be investigated. With this function it is easy to cancel out the factor $x - 3$ to give $x/(x + 3)$ which takes the value 0.5 when $x = 3$. The opportunity should not be missed to point out that the cancellation is only valid provided $x \neq 3$! In the other two examples given above it is possible to extract a factor $x - 3$ from the top and the bottom of the fraction but the manipulation involved is not so easy and may be outside the capability of the student.

Program 11.4 – To explore the limit of a function

```
 90 REM**PROGRAM 11.4**
100 REM**FUNCTION LIMIT**
110 DEF FNF(X) = (X*X - 3*X)/(X*X - 9)
120 PRINT "WHAT LIMITING VALUE FOR X?"
125 INPUT XL
130 PRINT "WHAT INITIAL VALUE FOR H?"
135 INPUT H
200 REM**CALCULATION SECTION**
210 X = XL + II
215 Y = FNF(X)
220 PRINT "X = ";X;" F(X) = ";Y
230 H = H/2
240 FOR J = 1 TO 1000 : NEXT J
250 GOTO 210
```

Below we give some other functions that the reader may like to investigate in the same way. Not all the functions converge as x approaches the stated value.

$$\frac{\sqrt{x^2+7}-4}{x^2-9} \quad \text{for} \quad x \to 3 \qquad\qquad \frac{x^2-6x+9}{\sqrt{x^2+7}-4} \quad \text{for} \quad x \to 3$$

$$\frac{\cos x}{1-\sin x} \quad \text{for} \quad x \to \frac{\pi}{2} \qquad\qquad \frac{\sqrt{x-1}}{\sqrt{x-1}} \quad \text{for} \quad x \to 1$$

$$\frac{\cos x - 1}{x^2} \quad \text{for} \quad x \to 0 \qquad\qquad \frac{\sin x}{x} \quad \text{for} \quad x \to 0$$

$$\frac{\sin(x^2)}{x} \quad \text{for} \quad x \to 0 \qquad \frac{\sqrt{x+1}-\sqrt{x}}{\sqrt{x}} \quad \text{for} \quad x \to 0$$

$$\frac{x^n - 1}{x - 1} \quad \text{for} \quad x \to 1 \quad \text{Try various values for } n.$$

$$\frac{\sin(x - \alpha)}{x^2 - \alpha^2} \quad \text{for} \quad x \to \alpha \quad \text{Try various values of } \alpha.$$

11.4 LIMIT OF A FUNCTION AS $x \to \infty$

Clearly it is possible to use the computer to get numerical experience of functions approaching a limit as x becomes large. At each step the value of x needs to be increased by some factor. In Program 11.5 this factor has been chosen as 10 and the function as $(x^2 + x - 10)/(3x^2 + 4)$. The resulting output indicates clearly the approach to the limit 1/3. Many similar examples can easily be constructed by the reader but perhaps these are not quite so obvious as most.

$$\frac{\sqrt{x+1}-1}{\sqrt{x}} \qquad \frac{\sqrt{x+1}+\sqrt{x}}{\sqrt{x}} \qquad \sqrt{x+1000}-x$$

Program 11.5 – To evaluate a function for increasing values of x

```
 90 REM**PROGRAM 11.5**
100 REM**FUNCTION LIMIT X-LARGE**
110 DEF FNF(X) = (X*X + X - 10)/(3*X*X + 4)
120 X = 1
200 REM**CALCULATION SECTION**
210 Y = FNF(X)
220 PRINT "X ="; X; "F(X) = "; Y
230 FOR J = 1 TO 1000 : NEXT J
240 X = 10*X
250 GOTO 210
```

11.5 THE TORTOISE AND THE HARE

This old paradox is worth using for the insight it can give into the summation of series. The tortoise challenged the hare to a race but the hare thought this was nonsense – of course he would win. However, the hare was finally persuaded and gave the tortoise a generous start. Now the hare runs much faster than the tortoise and soon reaches the place where the tortoise started. By this time the tortoise is a small distance ahead. The hare now runs to that position but again the tortoise must be ahead though only by a small amount. This argument may be repeated as many times as you like but the tortoise will still be a small amount in front, so the hare will never catch up the tortoise and cannot win the race.

Of course there is a flaw in the argument but the calculation of the times involved can help to reveal what is going on. Suppose the hare runs at 10 km/h and the tortiouse crawls at 1 km/h and let the hare give the tortoise a 1 km start.

The results may be tabulated:

Step no.	Time interval	Total time	Hare moves	Tortoise moves	Hare's total	Tortoise's total	Hare behind by
0	0.	0	0	0	0	1	1
1	0.1	0.1	1	0.1	1	1.1	0.1
2	0.01	0.11	0.1	0.01	1.1	1.11	0.01
3	0.001	0.111	0.01	0.001	1.11	1.111	0.001

A simple computer program can be written for this situation and run for as long as desired. The output line 210 can be altered to print other values if required, but in the example, it prints the three that demonstrate the convergence to a limiting total time that is still finite as the hare catches the tortoise up.

Program 11.6 – The tortoise and the hare

```
 90 REM**PROGRAM 11.6**
100 REM**HARE AND TORTOISE**
110 S=0 : TI=0 : TT=0 : HM=0 : TM=0 : HT=0 : TT=1 : HB=1
200 REM**CALCULATION SECTION**
210 PRINT "STEP NO. "; S; "TOTAL TIME"; TT;
    "HARE LAGS BY"; HB
220 S = S + 1
225 TI = HB/10
230 TT = TT + TI
235 HM = HB : TM = TI*1
240 HT = HT + HM : TT = TT + TM
245 HB = TT - HT
250 FOR J = 1 TO 1000 : NEXT J
255 GOTO 210
```

11.6 THE SUM OF AN INFINITE SERIES

The case of the tortoise and the hare introduces the subject of infinite series. It can come as a great surprise when the learner first discovers that an infinite sum can have a finite answer. The idea is not easy to accept at first, and some numerical experience using a computer can be of great value in the initial stages. It can also help to clarify ideas of radius of convergence in later work, and provides a basis for the development of convergence tests. We start with an example that may be derived in elementary algebra.

By applying the long division algorithm to the algebraic expression $1/(1-x)$ we get the following:

$$
\begin{array}{r|l}
 & 1 + x + x^2 + x^3 + \ldots \\
1-x & 1 \\
 & \underline{1-x} \\
 & \quad x \\
 & \quad \underline{x - x^2} \\
 & \qquad x^2 - x^3 \\
 & \qquad \underline{x^2 - x^3} \\
 & \qquad\quad \underline{x^3 - x^4} \\
 & \qquad\qquad x^4 \\
 & \qquad\qquad \ldots
\end{array}
$$

It is now clear that the process will never end; but does it have any meaning? To explore this we look at the succession of partial sums 1, $1 + x$, $1 + x + x^2$, $1 + x + x^2 + x^3$, etc. These may be computed by a program such as Program 11.7. We use N for the number of terms taken so far, U for the latest term and S for the sum of those terms. By experimenting with different values of x it will be found that the series does make sense in some cases. It may be worth asking what remainder is left if the long division process is stopped before the end. The convergence or otherwise of the partial sums of the series is apparent from the numerical evidence, and students will be able to guess the range of values of x for which it does converge. They will also see the slower rate of convergence as one gets near the ends of the range. It is also instructive to discuss the bahaviour when x is outside the range. If $x = 2$ the partial sums are 1, 3, 7, 15, ... and rapidly increase. With $x = -2$ they are 1, −1, 3, −5, 11, −21, ... and we get alternate signs as well as increasing size. The size of the remainder at each stage of the long division process can provide an explanation of the behaviour of the series.

Program 11.7 – To find the partial sums of the power series expansion of $1/(1-x)$

```
 90 REM**PROGRAM 11.7**
100 REM**PARTIAL SUMS OF SERIES FOR 1/(1-X)**
110 PRINT "WHAT VALUE FOR X";
115 INPUT X
120 PRINT "HOW MANY TERMS ";
125 INPUT T
130 U = 1 : S = 1 : N = 1
200 REM**CALCULATION SECTION**
210 PRINT "AFTER"; N; "TERMS THE SUM IS "; S
215 PRINT "LAST TERM = "; U
220 IF N = T THEN END
230 N = N + 1
235 U = U*X
240 S = S + U
250 GOTO 210
```

11.7 THE BINOMIAL SERIES

The expansion of $1/(1 - x)$ obtained by long division is a special case of the infinite power series obtained when the binomial theorem is applied to a negative, or non-integral positive, index. We have

$$(1+x)^p = 1 + \frac{p}{1}x + \frac{p(p-1)}{1.2}x^2 + \frac{p(p-1)(p-2)}{1.2.3}x^3 + \ldots \qquad (11.8)$$

where the infinitive series on the right may or may not converge to the value of the expression on the left depending on the value of x. If $p = -1$ and x is replaced by $-x$ we have the special case above. To compute the partial sums of the series in (11.8) we can evaluate each term from its predecessor then add it to the previous partial sum. Let the terms be u_0, u_1, u_2 etc. so that

$$u_0 = 1$$

$$u_1 = \frac{px}{1} = u_0 \times \frac{p}{1} \times x$$

$$u_2 = \frac{p(p-1)}{1.2}x^2 = u_1 \times \frac{p-1}{2} \times x$$

$$u_3 = \frac{p(p-1)(p-2)}{1.2.3}x^2 = u_2 \times \frac{p-2}{3} \times x$$

$$u_n = \frac{p(p-1)(p-2)\ldots(p-n+1)}{1.2.3\,.\,.\,.\,.\,.\,n}x^n = u_{n-1} \times \frac{p-n+1}{n} \times x\,. \qquad (11.9)$$

To proceed from one term to the next we need n, the number of the term, $p - n + 1$ which decreases by 1 at each stage, and x. In Program 11.8 the current value of $p - n + 1$ is in store M and this is reduced by 1 at each stage in line 235. Line 225 implements equation (11.9) and line 230 adds the new term to the previous partial sum.

Theoretically the binomial series converges for any x in the range $-1 < x < 1$ and any value of p. It is interesting to see how this emerges from experimentation with Program 11.8. The use of the program can lead to the conjecture that this is so, and can motivate the desire to prove it analytically.

Program 11.8 – To evaluate the partial sums of the binomial power series

```
 90 REM**PROGRAM 11.8**
100 REM**BINOMIAL POWER SERIES FOR (1 + x)↑P**
110 PRINT "WHAT VALUE FOR P ";
115 INPUT P
120 PRINT "WHAT VALUE FOR X";
```

```
125 INPUT X
130 S = 1 : M = P : N = 0 : U = 1
150 PRINT "HOW MANY TERMS";
155 INPUT T
200 REM**CALCULATION SECTION**
210 PRINT "N = "; N; "SUM = "; S
215 IF N >= T THEN END
220 N = N + 1
225 U = U*M*X/N
230 S = S + U
235 M = M - 1
240 GOTO 210
```

11.8 THE EXPONENTIAL SERIES

The power series for e^x is theoretically shown to converge for any value of x. The calculation of the partial sums for a variety of values can be quickly done with a modified version of Program 11.8 and gives some experience to underpin this idea. We have:

$$e^x = 1 + \frac{x}{1!} + \frac{x^2}{2!} + \frac{x^3}{3!} + \frac{x^4}{4!} + \ldots + \frac{x^n}{n!} + \ldots, \tag{11.10}$$

in which the nth term is

$$u_n = \frac{x^n}{n!} = u_{n-1} \times \frac{x}{n} \; . \tag{11.11}$$

The recurrence relation (11.11) is (11.9) with $p - n + 1$ omitted. We thus only need to change a few lines of Program 11.8 and the changes are given as Program 11.9.

Program 11.9 – Changes to Program 11.8 to produce the partial sums of the exponential series

```
 90 REM**PROGRAM 11.9**
100 REM**EXPONENTIAL SERIES**
130 S = 1 : N = 0 : U = 1
225 U = U*X/N
```

Omit lines 110, 115 and 235

11.9 THE LOGARITHMIC SERIES

The power series for the natural logarithm is:

$$\ln(1+x) = x - \frac{x^2}{2} + \frac{x^3}{3} - \frac{x^4}{4} + \frac{x^5}{5} - \frac{x^6}{6} + \ldots + (-1)^{n+1}\frac{x^n}{n} + \ldots \tag{11.12}$$

Each term cannot be directly computed from the one before, but in the program we can use store XN for the value of $(-1)^{n+1}x^n$ and use this to add the appropriate term to the partial sum. At each stage the new value for XN may be obtained from the old and in Program 11.10 it is line 225 that does this. Theoretically the series converges for $-1 < x \leqslant 1$ and a variety of values both inside and outside this range should be tried. It is valuable to see just how many terms are needed to achieve a desired accuracy in various cases. This can lead to a valuable discussion of the usefulness of this series for actually calculating a table of logarithms. It should be noted that the series diverges for $x > 1$ and yet $\ln(1+x)$ still has a value. Once again we have a power series where convergence is only obtained to certain values of the function, whereas the function is defined for a much wider range of values. This also happened with the binomial series.

Program 11.10 – The calculation of partial sums of the logarithmic series

```
 90 REM**PROGRAM 11.10**
100 REM**LOGARITHMIC SERIES**
120 PRINT "WHAT VALUE FOR X";
125 INPUT X
130 N = 1 : S = X : XN = X
150 PRINT "HOW MANY TERMS ";
155 INPUT T
200 REM**CALCULATION SECTION**
210 PRINT "N = "; N; "SUM = "; S
215 IF N >= T THEN END
220 N = N + 1
225 XN = -XN * X
230 S = S + XN/N
240 GOTO 210
```

The desire to be able to compute logarithms leads to the use of other series. In particular by putting $-x$ for x in (11.12) we have:

$$\ln(1-x) = -x - \frac{x^2}{2} - \frac{x^3}{3} - \frac{x^4}{4} - \ldots - \frac{x^n}{n} \ldots \tag{11.13}$$

and by subtracting (11.13) and (11.12) we get:

$$\ln\left(\frac{1+x}{1-x}\right) = \ln(1+x) - \ln(1-x) = 2x + 2\frac{x^3}{3} + 2\frac{x^5}{5} + 2\frac{x^7}{7} + \tag{11.14}$$

In modifying Program 11.10 to sum this series we have to decide how to number the terms since the nth term is now $2[(x^{2n-1})/(2n-1)]$. In Program 11.11 store D is used for the divisor in each term and XN now contains the current value of $2x^{2n-1}$. N is still used to count the number of terms taken.

Program 11.11 – Lines to be added or changed in Program 11.10 to sum the power series for $\ln[(1+x)/(1-x)]$

```
90  REM**PROGRAM 11.11**
100 REM**SERIES FOR LN((1 + X)/(1 – X))**
130 N = 1 : S = 2*X : XN = 2*X : D = 1
222 D = D + 2
225 XN = XN*X*X
230 S = S + XN/D
```

For both (11.12) and (11.14) convergence is for values of x within a limited range. A useful comparison of the two can be made by running the programs with values of x that lead to the same logarithm. For example $x = 0.5$ in (11.12) and $x = 0.2$ in (11.14) both lead to $\ln(1.5)$ and using these values in Programs 11.10 and 11.11 enable the different performance to be seen. The availability of the computer makes possible a wide exploration of the use of equation (11.14) since many values of x can be tried in a short time without the interest being killed by large amounts of tedious computation. For example with $x = 0.5$ we get $\ln 3$ and so can obtain $\ln 2$ from $\ln 3 - \ln 1.5$. It can be quite a challenge to devise values of x that lead to the natural logarithms of all the integers from 2 to 10.

11.10 THE SINE AND COSINE SERIES

The series for sine and cosine theoretically converge for all values of x, but unless x is small the partial sums do not at first appear to converge. The numerical evidence suggests that the series may diverge in some cases. This conflict between the students subjective judgement and the theory makes this a useful area for exploration and discussion, but it needs to be handled with care lest the wrong conclusions be drawn. The experience can give a useful warning to those who are inclined to place too much reliance on numerical evidence without developing the necessary analysis. Nowhere in this chapter do we suggest that numerical experience can replace a conventional analysis course, but we do believe that it can help conceptual development and illuminates what is for most a difficult area of mathematics.

For the computation of the partial sums of the sine series we can obtain the terms recursively. Equation (11.15) shows that $u_n = u_{n-1} \times [-x^2/((2n-2)(2n-1))]$ and this forms the basis for the calculation.

$$\sin(x) = x - \frac{x^3}{3!} + \frac{x^5}{5!} - \frac{x^7}{7!} + \ldots + \frac{(-1)^{n-1}x^{2n-1}}{(2n-1)!} \ldots \qquad (11.15)$$

Rather than using n in the calculation it is easier to keep a record of the current power of x and in Program 11.12 store D is used for this. To change one term into the next we multiply by x twice, change the sign, and divide by two successive values of D. For example, if the current value of u is $x^5/5!$ then $D = 5$. Line 222 makes $D = 6$ and u becomes $-(x^6/6!)$ then line 225 makes $D = 7$ and u becomes $-(x^7/7!)$ as required.

We have found it useful to suggest the use of some values of x for which the result is known. With $x = \pi/6$ convergence is rapid and the expected result of $\sin(\pi/6) = 0.5$ appears quickly. With $x = 25\pi/6$ the value should also be 0.5 but the behaviour of the series in this case is perhaps surprising to those who have not seen it before. It is important that enough terms are taken for the observer to see convergence to the same value in this case too.

Program 11.12 – To compute the partial sums of the sine series

```
 90 REM**PROGRAM 11.12**
100 REM**SINE SERIES**
120 PRINT "WHAT VALUE FOR X ";
125 INPUT X
130 N = 1 : U = X : S = X : D = 1
150 PRINT "HOW MANY TERMS ";
155 INPUT T
200 REM**CALCULATION SECTION**
210 PRINT "N = "; N; "SUM = "; S
215 IF N >= T THEN END
220 N = N + 1
222 D = D + 1 : U = -U*X/D
225 D = D + 1 : U = U*X/D
230 S = S + U
240 GOTO 210
```

The cosine series is similar and may be explored by a program like Program 11.12 with a few small changes. Equation (11.16) shows the same kind of recurrence relation between the terms and only the initial values need to be altered. Program 11.13 gives the changes to Program 11.12 and may be used in a similar way.

$$\cos(x) = 1 - \frac{x^2}{2!} + \frac{x^4}{4!} - \frac{x^6}{6!} + \frac{x^8}{8!} \cdots \qquad (11.16)$$

Program 11.13 – Changes to Program 11.12 for the cosine series

```
 90 REM**PROGRAM 11.13**
100 REM**COSINE SERIES**
130 N = 1 : U = 1 : S = 1 : D = 0
```

11.11 THE NATURAL LOGARITHM FUNCTION

One possible introduction to the natural logarithm function starts with the impossibility of finding a function of the form ax^p for $\int(1/x)\mathrm{d}x$. Clearly, from the graph of $1/x$, the area under the graph exists and the integral $\int_1^a(1/x)\mathrm{d}x$ is defined as $\ln(a)$. We have used a computer as a means of evaluating this integral for various values of a. Perhaps at the start it is best not to name this function but merely to discuss the need for such a function and ask what properties it possesses. We have found that many of the properties can be guessed from the numerical evidence obtained, and tested by evaluating it for other values of a. The trapezium rule (Program 13.5) with sufficient strips to give three- or four-figure accuracy has been found to be suitable for the evaluation. The suggestions made and tested by students include:

$$\begin{aligned}
\ln(a^2) &= 2\ln(a) \\
\ln(ab) &= \ln(a)+\ln(b) \\
\ln(a/b) &= \ln(a)-\ln(b) \\
\ln(a^3) &= 3\ln(a)
\end{aligned}$$

This last then leads to the conjecture that $\ln(a^p) = p\ln(a)$ and may be tested for other values of p. The question can also be posed: 'What value of 'a' results in $\ln(a) = 1$?'. Trial and error eventually produces a value of e that may be recognised by those who have met it before.

12

Introducing calculus

12.1 THE SLOPE OF THE GRAPH OF A FUNCTION

Early ideas about the slope of a graph of a continuous function are derived from drawing a variety of graphs. It is useful if this experience includes some calculation of slopes by dividing differences. The function $f(x) = 3x + 2$ includes the points $(2, 8)$ and $(5, 17)$ so that the slope of the line joining them is $(17-8)/(5-2) = 9/3 = 3$. This is of the form $[f(5) - f(2)]/(5 - 2)$ or $[f(2+h) - f(2)]/h$ when

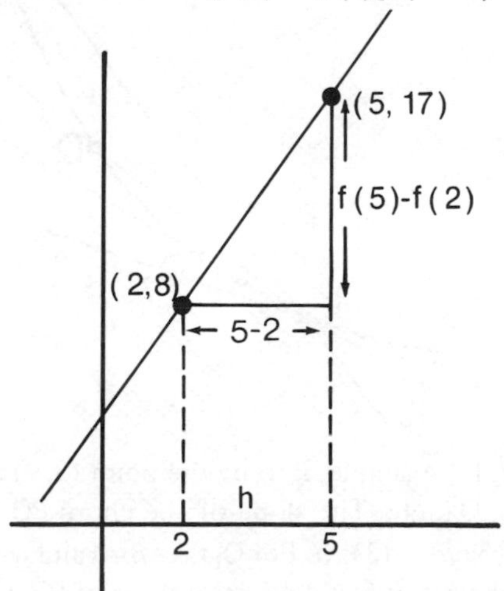

$h = 3$. The evaluation of this formula for various values of h leads always to the answer 3 and reinforces the idea that any segment of the straight line graph has the same slope. A number of such calculations leads to the idea that the slope of the graph of $f(x) = mx + c$ is measured by the number m irrespective of the value of c. Such calculations, and the generalisation that follows, form a valuable basis for the next step in developing the differential calculus, but cannot on their own lead to the idea that the slope is measured by another, related, function. That the derived function is a constant in this case acts against such an idea being formed rather than assisting it. There is need for some easily accessible way of getting similar numerical experience with non-linear functions so that a large number of cases can be examined quickly.

12.2 THE SLOPE OF A CURVE AT A GIVEN POINT

For the next part of our discussion we will use the function $f(x) = x^2$ although the methods can be applied to any function. The graph is a curve so the idea of the slope at a point needs some discussion. Through P we may draw straight lines such as PQ_1, PQ_2, etc. and all but one of them cut the curve again. PT is the only one not to cut the curve again and is the tangent at P. The slope of the tangent PT cannot be calculated directly from the formula, but the slopes of chords PQ_1, PQ_2, etc. can. Discussion of the diagram leads to the conclusion that if a succession of chords is taken, all with P as one end-point, and the other end-points closer and closer to P, then the calculated slopes will get nearer and nearer to the slope of PT. The slope of this tangent is then the slope of the curve at the point P.

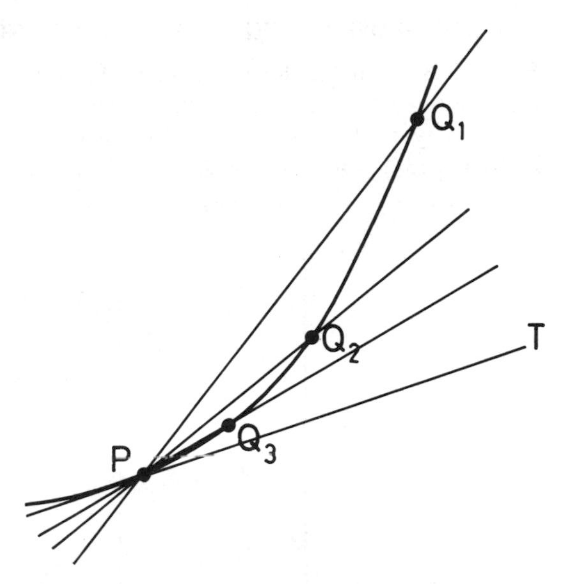

Take, for example, P to be the point (3, 9) and a succession of positions for Q as Q_1, Q_2, Q_3, etc. The slope of the chord PQ is $[f(3+h) - f(3)]/[(3+h) - 3]$, that is, $[f(3+h) - f(3)]/h$. For Q_1 take $h=1$ and we get a slope of $[(3+1)^2 - 3^2]/1 = 7$. For Q_2 take $h = 0.5$ and we get a slope of $[(3+0.5)^2 - 3^2]/0.5 = 6.5$.

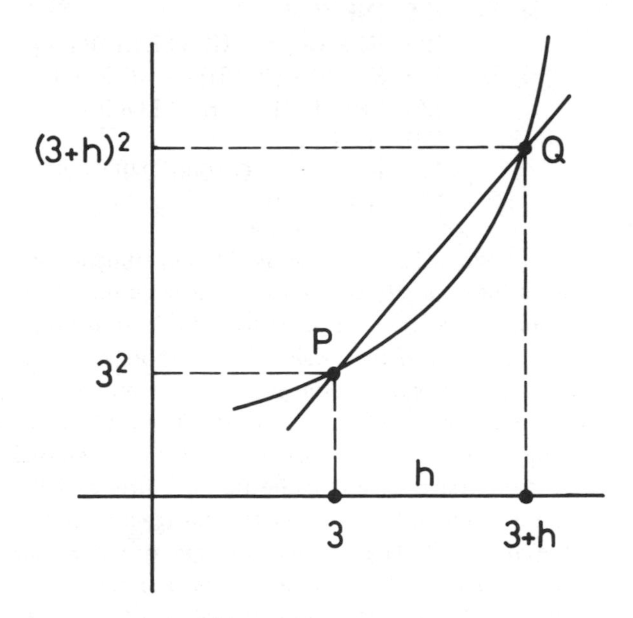

It is probably worth evaluating the slope for several values of h with a calculator and seeing the progression towards a limiting value, but a simple computer program can take the labour out of the calculation and leave the student free to look at the pattern in the results. The program can conveniently provide for the insertion of any function of the user's choice, and in Program 12.1 this may be done by changing line 110. The program provides for the user to choose the x-value of P and the initial value for h. The slope of the chord PQ is then evaluated for a sequence of values of h, each one half of the one before. The program does not have an END statement and execution should be terminated with the stop key. This is no real disadvantage, but if it is desired to allow ten steps, say, then a counter to do this can easily be inserted by including the lines:

```
140  C = 0
215  C = C + 1
250  IF C <= 10 THEN GOTO 210
```

Line 240 is simply a delay to prevent the values appearing on the screen too quickly. The value of 1000 in this line may be changed to vary the delay time.

Program 12.1 – To find the slopes of a succession of chords on the graph of a function, all having a common endpoint

```
 90  REM**PROGRAM 12.1**
100  REM**CHORD-SLOPE**
110  DEF FNF(X) = X*X
120  PRINT "WHAT FIXED X-VALUE ";
125  INPUT X
130  PRINT "WHAT INITIAL VALUE FOR H";
```

```
135 INPUT H
200 REM**CALCULATE SLOPE**
210 S = (FNF(X + H) - FNF(X))/H
220 PRINT "H = "; H; "SLOPE = "; S
230 H = H/2
240 FOR J = 1 TO 1000 : NEXT J
250 GOTO 210
```

Program 12.1 may be used to experiment with the situation in several ways. By using a variety of values of x a table may be built up giving the slope of the graph for each one. This activity, in itself, is establishing the idea that the slope is a function of the x-value. The authors have found that pupils experimenting with this program soon guess the formulae for the slope of graphs of such functions as x^2, $x^3/3$, x^3, $1/x$, etc. The activity can now be directed towards the tabulation of the functions used and the corresponding slope functions obtained. It is our experience that the rule for differentiating a polynomial emerges from this activity without the need for the teacher to state it. By using a microcomputer the necessarily large amount of numerical evidence can be quickly obtained and the learner's own ideas followed up. In addition, the fundamental idea used paves the way for the analytical derivation of the formula that may be met later as $f'(a) = \lim_{h \to 0} \{[f(a+h) - f(a)]/h\}$.

The student needs to be aware of the limitations on the accuracy of the computer when using Program 12.1. Because two nearly equal numbers are being subtracted when evaluating $f(a+h) - f(a)$ with small h, the error in the resulting number can swamp the number being calculated and lead to curious results. For example, taking $f(x) = \sqrt{x}$ at $x = 250\,000$ and using a microcomputer that works to about 9 figures, the value of

$$\lim_{h \to 0} \frac{f(250\,000 + h) - f(250\,000)}{h} \quad \text{is} \quad \frac{1}{2\sqrt{250\,000}} = 0.001$$

but the following sequence of values was obtained:

h	$\dfrac{f(250\,000+h)-f(250\,000)}{h}$
1	0.001 000 079
0.5	0.000 999 992
0.25	0.001 000 056
0.125	0.000 999 707
0.0625	0.000 999 965
0.031 25	0.001 000 479
0.015 625	0.000 993 878
0.007 812 5	0.000 995 934
0.003 906 3	0.000 987 879
0.001 953 1	0.000 993 013
0.000 976 6	0.001 009 464
0.000 030 5	0.004 959 106

Even from the start there is little sign of convergence, and the last value shows that by taking h too small the errors of computation corrupt the answer completely. Clearly the sequence should not be allowed to go too far. For further discussion of this point see Chapter 10.

12.3 THE AREA UNDER A GRAPH

There are many approaches to integration but, as with differentiation, all too often there is a rush into the development of techniques before there has been much experience of fundamental ideas. One approach starts with the idea of measuring the area under a graph. This links well with the wider use of integration for summation in a variety of applications. The idea is also easy to comprehend and, with a computer, easy to apply practically. We suggest the use of the trapezium rule to approximate the area since this requires nothing more than elementary mathematics. The area under the chord is given by $[f(a)+f(b)]/2 \times h$ where $h = b-a$. The smaller the value of h the smaller the error becomes so that in the second diagram, with two strips, the error (now the sum of two shaded pieces) is reduced. The value of h is now $(b-a)/2$ and the approximation to the area is

$$\left(\frac{f(a)+f(a+h)}{2}+\frac{f(a+h)+f(b)}{2}\right)\times h$$

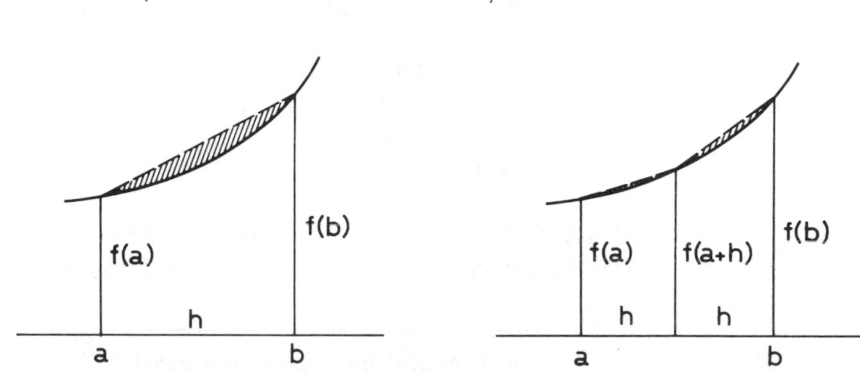

A computer program may be constructed to allow the number of strips to be increased (and hence the value of h reduced) successively, thus obtaining a sequence of approximations to the area. Although the trapezium rule is usually quoted in a form that amalgamates the two terms using the same value of $f(x)$ from adjacent strips, we favour the treatment of each strip separately so that the program is more easily understood. On a modern microcomputer the difference in computational time is not serious. In Program 12.2 the function is defined in line 110 and only this line need be retyped to change to a different function. The program runs continuously until interrupted from the keyboard. There is no need for a delay between one value of N and the next since, after the first few cases, the time taken becomes appreciable and quite enough for the output to be studied. This point is itself worthy of discussion leading to an explanation of why each new case takes as long as all the preceding cases together.

As with the chord-slope, Program 12.1, the calculation errors can become important if excessively small values of h are used, though for a rather different reason. Since each x-value is obtained from the one before by adding h, errors can accumulate by the end of the range if the number of strips is large.

Program 12.2 – To calculate the area under the graph of* $3X^2$ *between* X = XL *and* X = XH *using the trapezium rule

```
 90 REM**PROGRAM 12.2**
100 REM**TRAPEZIUM RULE**
110 DEF FNF(X) = 3*X*X
200 REM**INPUT SECTION**
210 PRINT "WHAT LOWEST X";
215 INPUT XL
220 PRINT "WHAT HIGHEST X";
225 INPUT XH
230 N = 2
300 REM**CALCULATION SECTION**
305 A = 0 : X = XL
310 H = (XH - XL)/N
320 FOR J = 1 TO N
325    A = A + (FNF(X) + FNF(X + H))/2*H
330    X = X + H
335 NEXT J
350 PRINT "WITH"; N; "STRIPS AREA IS"; A
355 N = N*2
360 GOTO 300
```

Using $f(x) = x^2$, XL = 0 and XH = 3 the exact value of the area is 9. Program 12.2 with this function gave the following results:

Number of strips	Calculated area
2	10.125
3	9.5
6	9.125
10	9.045
100	9.000 449 98
1 000	9.000 004 24
10 000	9.000 001 91
100 000	9.000 028 51

There is clear indication of convergence towards the correct value but, though theoretically 100 000 strips should give a better value still, in practice it gives a worse one.

The program may be used to investigate the area under the graph of, for example, $f(x) = x^2$. Various ranges can be tried but, at some point, it is valuable to accumulate results for a series of intervals all having the lower value as XL = 0 and varying the higher value XH. A table of values of area against XH can lead to guessing, and testing, of a formula. If no formula is suggested then using the function $f(x) = 3x^2$ will most certainly produce a result. In this way the numerical exploration suggests that there are related functions; for example, with $f(x) = x^2$ then the function $F(x) = x^3/3$ measures the area under the graph of $f(x)$ from 0 to each x-value.

A further investigation then suggests itself, to find the area between two values of x neither of which is 0. The result can lead to the formula $F(b) - F(a)$ and thus to the evaluation of a definite integral.

In all of this work no knowledge of the differential calculus is needed and the integral calculus may thus be introduced first.

13

Numerical techniques

In this chapter we include some of the more commonly used numerical techniques that may be included in the school curriculum. All are widely represented in the literature on numerical analysis, and for a more detailed description of the methods and their derivation the reader should refer elsewhere. Here we shall give a simple introduction to each technique and indicate how it may be successfully programmed for a microcomputer.

13.1 THE SOLUTION OF EQUATIONS OF ONE VARIABLE BY INTERVAL BISECTION

In Chapter 3 we indicate a method of solution that can arise naturally in a trial-and-error approach to equation solving. Two values of the variable are found one of which is clearly too large and the other too small so that the solution is trapped between them. A value between the two is now tested to see if it is too large or too small and replaces one of the previous values to give a smaller interval containing the solution. The last step is now repeated until the interval is small enough to ensure that we know the solution to the required accuracy. In Chapter 2 the algorithm for finding a particular word in the dictionary by questions with yes or no answers indicated just how rapidly such a chopping of an interval can reduce its size.

To show the steps in the interval bisection method we will use as an example the equation:

$$x^3 - 6x^2 - 11x + 50 = 0 \qquad (13.1)$$

First we need to locate the solutions and this may be done by evaluating the function

$$f(x) = x^3 - 6x^2 - 11x + 50$$

for values of x in some suitable range. Program 3.1 given in the chapter on functions and equations is very suitable for this. Suppose we decide to explore the range 0 to 12 in steps of 1 then we obtain a table of values of the function:

x	0	1	2	3	4	5	6	7
$f(x)$	50	34	12	−10	−26	−30	−16	22

x	8	9	10	11	12
$f(x)$	90	194	340	534	782

These values may be represented on a graph.

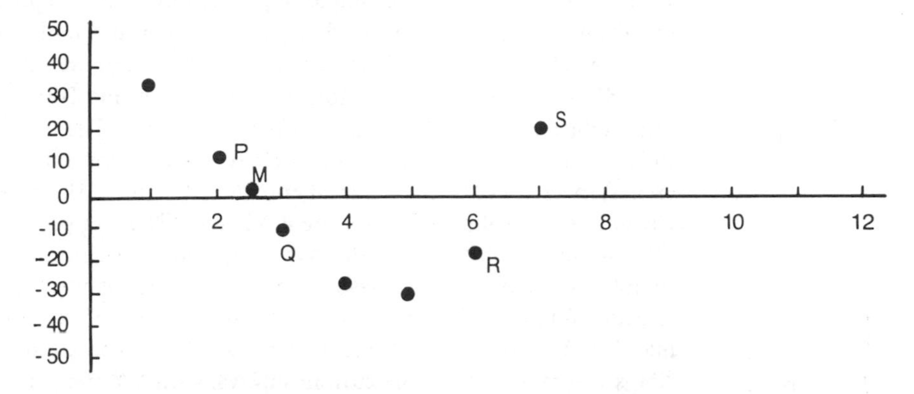

We now make the assumption (reasonable in this case) that the function is continuous and therefore the graph from P to Q crosses the axis somewhere between the two points and also another crossing between R and S. This shows that there is one solution with $2 < x < 3$ and another with $6 < x < 7$.

The next step is to improve the accuracy by evaluating the function at other values. In Chapter 3 we suggested dividing the interval into small steps and thus finding a smaller interval. Then for the 'open box' problem we describe an algorithm in which the mid-point of the interval is used each time, and it is this method that we use here; hence the method is one of successively bisecting the interval.

For the solution between $x = 2$ and $x = 3$ the mid-point is 2.5 and $f(2.5) = 0.625$. Since this is positive we now know that $2.5 < x < 3$, that is, we have replaced the lower end of the interval by the mid-value. In the diagram M replaces P and we know that the graph crosses the axis between M and Q. The procedure may now be applied again and we find that $f(2.75) = -4.828$ which is negative, so this time we replace the upper end of the interval to get $2.5 < x < 2.75$.

Each application of the process divides the interval by 2 so that 10 applications divide it by $2^{10} \approx 1000$. Thus 10 applications improve the accuracy of the result by about three decimal places. What is more, we have a method that gives an upper and lower bound on the solution.

A computer program to carry out the above process would consist of the following parts:

(1) A line defining the function $f(x)$.
(2) A section allowing the user to select the range of x-values for the search for a solution, the step size, and the size to which the interval containing the solution is to be reduced.
(3) To search through the range of values for x evaluating the function at each step. Whenever two successive values of $f(x)$ are of opposite sign, pass control to (4) below.
(4) Reduce the size of the interval containing the solution by successively evaluating $f(x)$ at the mid-point and replacing one or other end point according to the sign of this value of $f(x)$. Repeat until the interval size is small enough, then print the solution and return to 3.

Example Program 13.1 is a simple implementation of this process. The function definition in line 110 may be changed to solve a different equation. The range to be searched is XL to XH with a step size XS and these values are input in line 220. SI is set to the desired solution interval in line 240. For a four-decimal-place solution this would be given the value 0.0001. The search is carried out by the loop in lines 320 to 340. YL is set to the function value at the beginning of each step and YH to the value at the end (that is, at the current x-value). The test in line 330 calls the subroutine if YL and YH are opposite in sign.

The interval bisection process is carried out in the subroutine from lines 400 to 460, and may be called several times from the main program, once for each solution found. The x and $f(x)$ values at each end of the step are first copied into XP, YP, XQ and YQ and it is these stores that are used for the process. XM is the mid-point of the current interval with YM the corresponding function value. The test in 425 examines YP and YM. If they are of opposite sign then YP*YM < 0 and the solution lies between XP and XM so that XM becomes the

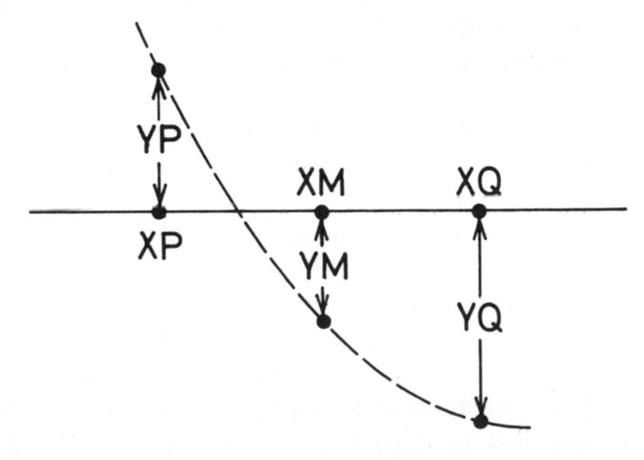

Diagram of the case where YP and YM are of opposite signs.

new XQ, otherwise XM becomes the new XP. Line 440 tests the interval size and if it is still greater than the desired interval SI the process is repeated from line 415. The final part of the subroutine prints an interval containing the root and counts the number of roots found in counter C.

When using Program 13.1 to investigate the solution of equations in this way it is valuable for the steps towards the solution to be seen. An extra PRINT statement may be inserted as line 412 to output the values of XP and XQ each time a new interval is calculated. Examination of the resulting lists of values can be quite instructive.

In certain special cases Program 13.1 will fail to work correctly or miss some solutions. If the steps of the search are so large that two solutions lie in the same step then no change of sign of $f(x)$ will be detected and those solutions will not be found. When one of the step points is itself a solution the program still works, but the result is given as two intervals with no indication that the solution is at the end of those intervals. A small modification to the program can be made by inserting a line 327 to test for YH = 0 and print the exact solution. Further complications arise if the function is discontinuous and the program should not be used for a range that contains a discontinuity. To illustrate this the reader may like to try solving the equation

$$\frac{x^2 - 7x + 8}{x - 4.21} = 0$$

where the function clearly has a discontinuity at $x = 4.21$ but also some solutions in the range 0 to 10.

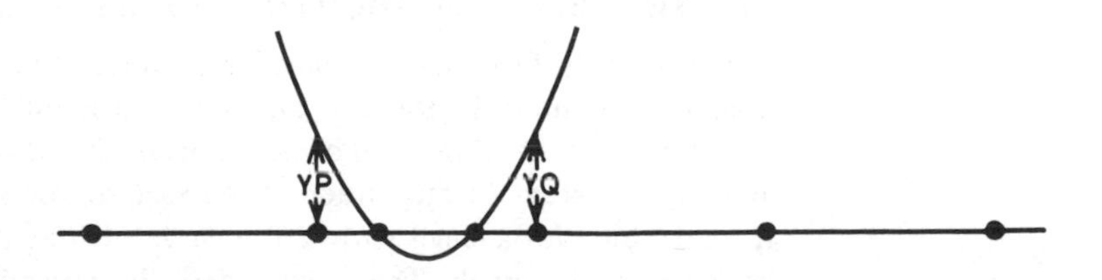

Diagram showing how a large step can miss a pair of solutions.

Program 13.1 – To solve the equation $f(x) = 0$ by interval bisection

```
 90 REM**PROGRAM 13.1**
100 REM**SOLVE EQUATIONS BY**
101 REM**INTERVAL BISECTION**
110 DEF FNF(X) = X*X*X - 6*X*X - 11*X + 50
200 REM**INPUT RANGE, STEP SIZE AND ACCURACY**
210 PRINT "WHAT LOWEST-X, HIGHEST-X, AND"
211 PRINT "STEP SIZE";
220 INPUT XL, XH, XS
230 PRINT "WHAT SIZE INTERVAL FOR THE"
231 PRINT "SOLUTION";
```

```
240 INPUT SI
300 REM**SEARCH RANGE FOR SOLUTIONS**
305 C=0
310 YL = FNF(XL)
320 FOR X = XL + XS TO XH STEP XS
325    YH = FNF(X)
330    IF YL*YH <= 0 THEN GOSUB 400
335    YL = YH
340 NEXT X
345 PRINT "THERE ARE"; C; "SOLUTIONS IN"
346 PRINT "THE RANGE"; XL; "TO"; XH
350 END
400 REM**INTERVAL BISECTION SUBROUTINE**
410 XP = X - XS : YP = YL : XQ = X : YQ = YH
415 XM = (XP + XQ)/2
420 YM = FNF(XM)
425 IF YP*YM < 0 THEN GOTO 435
430 YP = YM : XP = XM : GOTO 440
435 YQ = YM : XQ = XM
440 IF ABS(XQ - XP) > SI THEN GOTO 415
450 PRINT "THERE IS A SOLUTION BETWEEN"
451 PRINT XP;" AND "; XQ
455 C = C + 1
460 RETURN
```

13.2 THE SECANT METHOD (Some books refer to this as '*regula falsi*')

One intuitively feels that it is possible to move towards the solution much faster than in the interval bisection method if account is taken of the size of the function values at either end of the interval. This is particularly so if YP and YQ are very different in magnitude. The exact solution is where the graph crosses the axis, but, for a small part of the curve such as P to Q, the straight line will be close to the graph. The point where the straight line PQ cuts the axis will therefore apparently be a much better choice for XM than the mid-point. By similar triangles, remembering that YQ is of opposite sign from YP we have

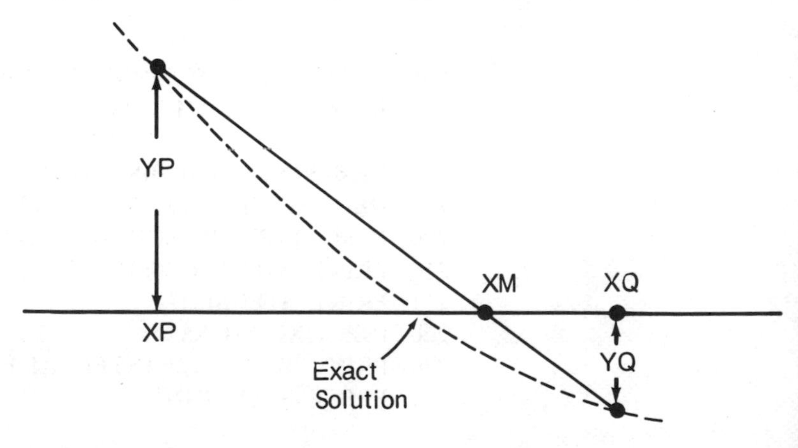

$$\frac{\text{XM} - \text{XP}}{\text{YP}} = \frac{\text{XQ} - \text{XM}}{-\text{YQ}}$$

which gives

$$\text{XM} = \frac{\text{XQ.YP} - \text{XP.YQ}}{\text{YP} - \text{YQ}} \tag{13.2}$$

and this could be used to find XM in line 415 of Program 13.1. A snag can arise when we try this improvement, however. In the case shown in the diagram, where the graph is concave upwards, the solution lies between XP and XM and so XM becomes the new value of XQ. The new interval is not so small as with the interval bisection process. Further applications of equation (13.2) result in XQ approaching the solution, but the interval remains large and can no longer be used as a criterion for halting the process, or as a test of the accuracy of the solution. Some alternative criterion for stopping the process will be needed such as requiring $f(\text{XM})$ to be less than some predetermined small value. This cannot be related so easily to a stated accuracy of the solution. Alternatively we may stop when successive value of XM change by a sufficiently small quantity at each step, however a change of less than 0.0001, for example, does not guarantee a solution to four decimal places since the sum of several changes of this magnitude could affect the fourth decimal place.

Changes to Program 13.1 to apply the secant method are given as Program 13.2. As with the interval bisection method, it can be useful to insert an extra PRINT statement as line 412 to output XP and XQ at each stage in obtaining a solution. Comparison with the interval bisection method can show the advantages and disadvantages of each.

Program 13.2 – Changes to Program 13.1 to apply the secant method

```
90  REM**PROGRAM 13.2**
230 PRINT "HOW SMALL SHOULD F(X) BE FOR"
231 PRINT "AN ACCEPTABLE SOLUTION";
415 XM = (XQ*YP - XP*YQ)/(YP - YQ)
440 IF ABS(YM) > SI THEN GOTO 415
450 PRINT "SOLUTION IS"; XM
451 PRINT "F(X) = "; YM
```

13.3 THE NEWTON–RAPHSON METHOD

In the secant method the assumption was made that for a small interval the graph of the function will be so nearly straight that using a straight line to join two points will lead to a good estimate of the solution. For this purpose two points on the graph were used; however, if we know the direction of the line then one point would be enough. Suppose we know that x_0 is near to the solution. P is the point on the graph with x-co-ordinate x_0 and we wish to choose a straight line through P that approximates the graph in some way. The line must be one whose direction can easily be calculated. The assumption that for a small interval the graph is nearly straight leads to the tangent to the graph being a

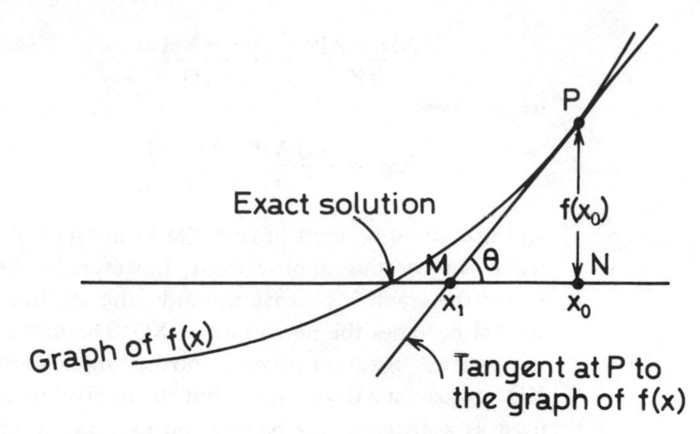

suitable choice. If this tangent at P cuts the axis at x_1 then we can calculate x_1 and it will be a better approximation to the solution. Now PN/MN = $\tan\theta$ so that $f(x_0)/(x_0-x_1) = \tan\theta$ or $x_0-x_1 = f(x_0)/\tan\theta$. However, $\tan\theta$, the slope of the tangent, is the value of the derivative of $f(x)$ evaluated at x_0, that is, $f'(x_0)$, so

$$x_0 - x_1 = \frac{f(x_0)}{f'(x_0)}$$

or

$$x_1 = x_0 - \frac{f(x_0)}{f'(x_0)} \tag{13.3}$$

Equation (13.3) allows us to calculate x_1 from x_0. We may now use x_1 as the approximation to the solution and apply the same formula to obtain x_2, a better approximation still. Equation (13.3) thus forms the basis for an iterative process that may be restated as:

$$x_{n+1} = x_n - \frac{f(x_n)}{f'(x_n)} \tag{13.4}$$

For our example equation (13.1) we have

$$f(x) = x^3 - 6x^2 - 11x + 50$$

so that

$$f'(x) = 3x^2 - 12x - 11 \ .$$

In writing a program to implement the Newton–Raphson method we must decide just how the starting value x_0 is to be obtained. It is simplest to let the user select it and is our choice in Program 13.3. An alternative is to use some search procedure such as that used in the interval bisection Program 13.1. The criterion for stopping the iterative process must also be decided, and in our example we test for a suitably small change in values. This is determined by the small number E that is selected by the user in line 225. The function is defined in line 110 and its derivative, called FND(X), in line 115. Only these two lines need be changed to solve a different equation. Line 325 causes all the iterates to be printed and this is useful to show how well the method performs in various cases. If only the final value is required then this line could be renumbered 335.

Program 13.3 – Solution of an equation by Newton–Raphson iteration

```
 90 REM**PROGRAM 13.3**
100 REM NEWTON–RAPHSON ITERATION**
110 DEF FNF(X) = X*X*X - 6*X*X - 11*X + 50
115 DEF FND(X) = 3*X*X - 12*X - 11
200 REM**INPUT STARTING VALUE**
201 REM**AND STOPPING CONDITION**
210 PRINT "WHAT STARTING X";
215 INPUT X
220 PRINT "HOW SMALL MUST CHANGE IN X BE FOR"
221 PRINT "ITERATION TO BE STOPPED";
225 INPUT E
300 REM**ITERATION**
305 C = 0
310 XP = X
315 X = XP - FNF(XP)/FND(XP)
320 C = C + 1
325 PRINT "AFTER"; C; "ITERATIONS X = "; X
330 IF ABS(X - XP) > E THEN GOTO 310
```

The Newton–Raphson method performs very well and usually gives rapid convegence to the solution. As a rough guide, once we are near the solution we usually get a doubling of the number of decimal places that are correct with each application of the process. It can run into difficulties where there are two solutions close together, or there is a turning point of the function near the solution.

The example equation given in Program 13.3 has three roots and exploring different starting values can lead to finding all three. The exploration should include various starting values between 4 and 5, and it can be instructive to try to explain the behaviour of the method for different values in this range. Reference to the graph of the function should enable the student to interpret the output from the program.

The good performance of the Newton–Raphson method tempts one to use this one straight away in preference to the earlier methods of this chapter. It is our belief however, that there is considerable merit in a method that can be developed, with guidance, from the student's spontaneous reactions to a problem. We have been able to do this with interval bisection but doubt whether Newton–Raphson would be 'discovered' in this way. Also, for students with no working knowledge of calculus it cannot be applied. The interval bisection method can be introduced much earlier in a mathematics course and gives good experience of an iterative process that is easily understood.

13.4 SETS OF SIMULTANEOUS LINEAR EQUATIONS

Often in a school course in mathematics simultaneous equations are limited to two equations in two variables such as

$$\begin{aligned} 4x + 3y &= 23 \\ 5x + 7y &= 45 \end{aligned} \tag{13.5}$$

and in most examples they are linear. Various methods of solution maybe explored, and often pupils are encouraged to look at the numbers and choose an approach that keeps the arithmetic as simple as possible. To a computer it does not matter whether the numbers are simple or not, but choosing a method to fit the particular numbers is difficult to program. It is better to have some systematic procedure that will apply in all cases (if possible).

The extension to three or more equations is seldom met in the usual school course, yet it would be sensible to have a method that readily extends to any number of equations. Amongst the methods of solution of equations (13.5) one of the most popular is to multiply the top equation by 5 and the bottom one by 4 to give

$$\begin{aligned} 20x + 15y &= 115 \\ 20x + 28y &= 180 \end{aligned}$$

The next step is to subtract, possibly as top from bottom to keep things positive, to get

$$13y = 65$$

then

$$y = 5 \ .$$

A decision has now to be taken as to which equation to use to find x. Going back to the original top equation of (13.5) and substituting for y we get

$$4x + 15 = 23$$

thus

$$x = 2 \ .$$

Extension of this approach to three equations such as

$$\begin{aligned} 2x - 3y + 4z &= 11 \\ 3x + 2.5y + 5z &= 21.5 \\ 5x - 11y + 8z &= 20 \end{aligned} \tag{13.6}$$

requires some care in keeping track of the equations to be used. The first stage uses the equations in pairs to eliminate x and yields two equations in y and z. These are in turn solved and the values used in one of the original equations to get x.

For computer solution a modification of this approach that is systematic can be used and the whole system is retained at each stage. There are two stages to the method. In the first stage at each step a multiple of one equation is added to another so that one term is reduced to zero. For equations (13.5) m times the top equation is added to the bottom equation to produce a zero x-term. The multiplier comes from the x-coefficients and is given by $m = -(5/4)$.

$$4x + 3y = 23$$

$$\times -\frac{5}{4}$$

$$0x + 3.25y = 16.25 \ .$$

For such a simple system this completes the first stage. The second stage consists of substituting to find the values of the variables. The second equation now gives $y = 16.25/3.25 = 5$ and this may be used in the first to find x. From the first $x = (23 - 3y)/4$ so

$$x = \frac{23 - (3 \times 5)}{4} = 2 \ .$$

In writing out the steps there is no need to write the variables if the coefficients are written as a matirx. For the set of three equations (13.6) the calculation proceeds as follows:

$$\begin{matrix} \\ m = -\frac{3}{2} \\ m = -\frac{5}{2} \end{matrix} \begin{bmatrix} 2 & -3 & 4 \\ 3 & 2.5 & 5 \\ 5 & -11 & 8 \end{bmatrix} \quad \begin{bmatrix} 11 \\ 21.5 \\ 20 \end{bmatrix}$$

$$\begin{matrix} \\ \\ m = -\frac{-3.5}{7} \end{matrix} \begin{bmatrix} 2 & -3 & 4 \\ 0 & 7 & -1 \\ 0 & -3.5 & -2 \end{bmatrix} \quad \begin{bmatrix} 11 \\ 5 \\ -7.5 \end{bmatrix}$$

$$\begin{bmatrix} 2 & -3 & 4 \\ 0 & 7 & -1 \\ 0 & 0 & -2.5 \end{bmatrix} \quad \begin{bmatrix} 11 \\ 5 \\ -5 \end{bmatrix}$$

This completes the first stage in which a triangle of zeros has been systematically created below the leading diagonal of the matrix. There is no restriction on the number of equations and it will work in the same way for any number. At the end of the stage there are still the same number of equations as at the start so we can see that we have an equivalent system, but one in which it is easy to find the values of the variables. The creation of the triangle of zeros is called Gaussian-elimination. In the second stage the equations are used in reverse order to calculate the unknowns. This is called back substitution.

$$-2.5z = -5 \qquad z = \frac{-5}{-2.5} = 2$$

$$7y - z = 5 \qquad y = \frac{5+z}{7} = 1$$

$$2x - 3y + 4z = 11 \qquad x = \frac{11 + 3y - 4z}{2} = 3 \ .$$

A computer program to use Gaussian-elimination to solve linear simultaneous equations will have the following sections:

1. Read the coefficients of the equations into a matrix and the constants into a corresponding vector.
2. Add multiples of rows of the matrix to other rows to create zeros below the leading diagonal at the same time adjusting the vector of constants correspondingly. Zeros are created in one column at a time starting from the first column.
3. Use back substitution to calculate the unknowns.
4. Output the solutions.

Program 13.4 is a simple and straightforward implementation of these sections. The coefficients of the equations are contained in a data section starting at line 1000. Only these lines need to be changed to solve a different set of equations. Line 115 sets N to the number of equations (and unknowns) in the system and reads this from line 1005. In our example N = 3. Line 120 then sets the dimensions of the matrices to the required size. A(N, N) is for the coefficients, B(N) for the constants and X(N) for the solutions. The nested loops in lines 130 to 155 read the coefficients into the matrix. J is the number of the equation being read and K the coefficient in that equation. The final line in the J-loop is to read the constant into B(J). Each DATA line from 1010 onwards corresponds to one equation. Thus when J = 3 in our example, K counts 1 to 3 and the numbers 5, −11, 8 are read into the third row of matrix A, the values being obtained from DATA line 1030. Finally the constant 20 is read into B(3).

The elimination section from lines 200 to 250 also consists of nested loops. The value of K is the column of the matrix where zeroes are currently being created. For each K value J counts through the equations in turn starting with equation K + 1. In the J-loop, lines 215 to 245, the first task is to calculate the multiplier M for that equation. Another counter L is then used (lines 225 to 235) to count through the terms of the Jth equation and add M times the corresponding term of the Kth equation to each. Finally in line 240 the constants are dealt with similarly.

An example may help to make this clear. The diagram shows a 5 by 5 matrix where entries that are still non-zero are shown by asterisks. It has reached the stage where elimination is proceeding in the second column, so K = 2. The next equation to be dealt with is the fourth, so $J = 4$. M will be calculated from the two underlined values and is $-(q/p)$, then L counts from 2 to 5 and M times the Lth term in equation 2 is added to the corresponding term in equation 4. Finally M times $B(2)$ is added to $B(4)$.

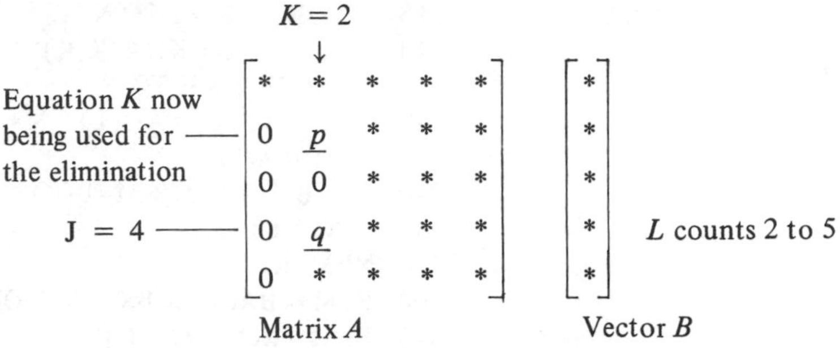

$K = 2$

↓

Equation K now being used for the elimination —— row 2; $J = 4$ —— row 4; L counts 2 to 5

$$\underbrace{\begin{bmatrix} * & * & * & * & * \\ 0 & \underline{p} & * & * & * \\ 0 & 0 & * & * & * \\ 0 & \underline{q} & * & * & * \\ 0 & * & * & * & * \end{bmatrix}}_{\text{Matrix } A} \quad \underbrace{\begin{bmatrix} * \\ * \\ * \\ * \\ * \end{bmatrix}}_{\text{Vector } B}$$

Diagram of the matrix part-way through the elimination in a case where $N = 5$.

The back substitution section starts with the calculation of X(N) in line 310. The loop in lines 315 to 345 now calculates the remaining solutions in reverse order. The value of J indicates the solution currently being calculated, and the calculation starts in line 320 with S being set to the constant B(J) from the Jth equation. The K-loop, lines 325 to 335, now subtracts from S the terms in the Jth equation involving the X's that have already been found. The final step in line 340 is to divide the result by the coefficient of X(J) to obtain the value of X(J). All that remains is to output the solutions in lines 410 to 420.

The simple Gaussian elimination process described above has defects that affect its performance in certain cases. If a zero value should occur on the leading diagonal at any stage in the process then M cannot be calculated and the process will stop. Near zero values can cause rapid error build up and in extreme cases completely corrupted solutions may result. It is better to keep the leading diagonal terms as large as possible and this may be achieved in part by exchanging equations as appropriate. The process that does this is called 'pivoting'. Gaussian elimination with pivoting is a widely used method of solving sets of linear simultaneous equations.

Program 13.4 – Solution of a set of linear simultaneous equations by Gaussian elimination

```
 90 REM**PROGRAM 13.4**
100 REM**SOLUTION OF SIMULTANEOUS**
101 REM**EQUATIONS BY GAUSSIAN**
102 REM**ELIMINATION**
110 REM**INPUT MATRIX AND VECTOR**
115 READ N
120 DIM A(N,N), B(N), X(N)
130 FOR J = 1 TO N
135    FOR K = 1 TO N
140       READ A(J,K)
145    NEXT K
150    READ B(J)
155 NEXT J
200 REM**ELIMINATION**
210 FOR K = 1 TO N
```

```
 215      FOR J = K + 1 TO N
 220        M = -A(J,K)/A(K,K)
 225        FOR L = K TO N
 230          A(J, L) = A(J, L) + M * A(K, L)
 235        NEXT L
 240        B(J) = B(J) + M * B(K)
 245      NEXT J
 250    NEXT K
 300    REM**BACK SUBSTITUTION**
 310    X(N) = B(N)/A(N,N)
 315    FOR J = N - 1 TO 1 STEP -1
 320      S = B(J)
 325      FOR K = J + 1 TO N
 330        S = S - A(J,K)*X(K)
 335      NEXT K
 340      X(J) = S/A(J,J)
 345    NEXT J
 400    REM**OUTPUT SOLUTIONS**
 410    FOR J = 1 TO N
 415      PRINT "X("; J;") = "; X(J)
 420    NEXT J
1000    REM**DATA SECTION**
1005    DATA 3 : REM*NUMBER OF EQUATIONS*
1010    DATA 2, -3,4, 11
1020    DATA 3, 2.5, 5, 21.5
1030    DATA 5, -11, 8, 20
```

13.5 NUMERICAL INTEGRATION TO FIND THE AREA UNDER THE GRAPH OF A FUNCTION – TRAPEZIUM RULE

(A different version of the program given in this section will be found in Chapter 12.)

The definite integral $\int_a^b f(x)\mathrm{d}x$ can be interpreted as the area under the graph of $f(x)$ between $x = a$ and $x = b$. There are many ways of approximating such an area, but one of the most common in schools is the trapezium rule that is

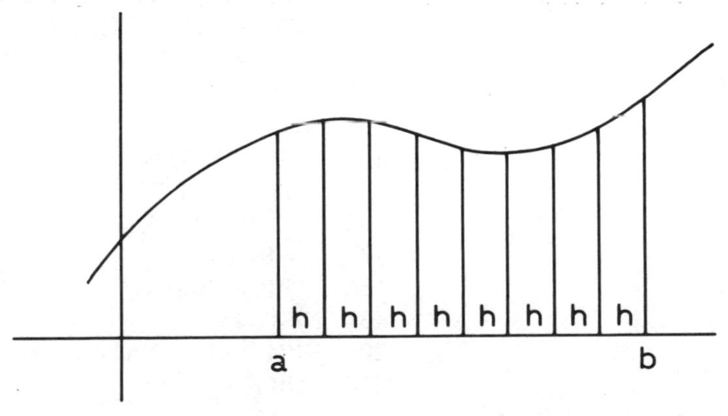

often met in quite elementary work. Essentially the method is to regard the required area as composed of a series of strips parallel to the y-axis. Each strip has a curved top boundary but provided the strips are not too wide we may replace this by a straight line and still have a reasonable approximation to the true area. If there are n strips of equal width h then $h = (b-a)/n$. The rth strip has x-values $a + (r-1)h$ and $a + rh$ at its ends so the heights of the two sides are $f(a+(r-1)h)$ and $f(a+rh)$. The area is thus

$$\frac{1}{2}(f(a+(r-1)h) + f(a+rh)) \times h.$$

Adding all the strips gives the formula

$$\begin{aligned}\text{Area} &= \frac{1}{2}(f(a) + f(a+h)) \times h + \frac{1}{2}(f(a+h) + f(a+2h)) \times h + \dots \\ &\quad + \frac{1}{2}(f(a+(n-1)h) + f(a+nh) \times h \\ &= \left(\frac{1}{2}f(a) + f(a+h) + f(a+2h) + f(a+3h) + \dots\right. \\ &\quad \left. + f(a+(n-1)h) + \frac{1}{2}f(a+nh)\right) \times h\ . \qquad (13.7)\end{aligned}$$

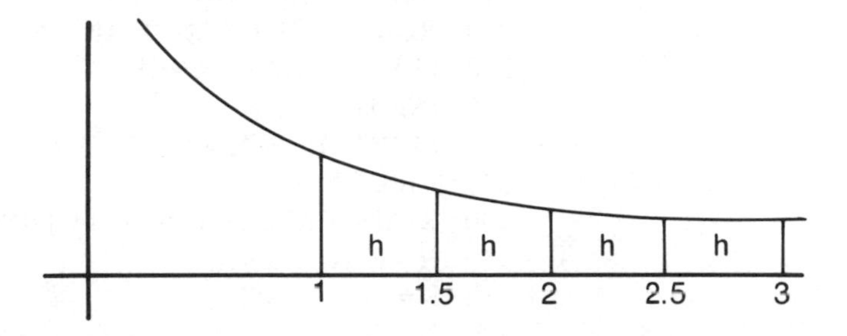

As an example consider the area under the graph of $1/(x+2)^2$ between $x = 1$ and $x = 3$. We choose to have 4 strips so $n = 4$ and $h = (3-1)/4 = 0.5$. Next we find S the sum of the required function values.

$$\begin{aligned}S &= \frac{1}{2}f(1) + f(1.5) + f(2) + f(2.5) + \frac{1}{2}f(3) \\ &= \frac{1}{2} \times 0.11111 + 0.08163 + 0.0625 + 0.04938 + \frac{1}{2} \times 0.04 \\ &= 0.26907\end{aligned}$$

The approximation to the area is A and

$$\begin{aligned}A &= S \times h \\ &= 0.26907 \times 0.5 \\ &= 0.13453\end{aligned}$$

For this function we may calculate the exact area by integration

$$\int_1^3 \frac{1}{(x+2)^2} = \left[-(x+2)^{-1}\right]_1^3 = -\frac{1}{5}+\frac{1}{3} = 0.13333$$

so even with only four strips the approximation to the area has an error of only about 1%. It is a valuable experience to try various values for n to see how changing the number of strips affects the accuracy. Program 13.5 may be used for this.

The function is defined in line 110 and can easily be changed by retyping just one line. The calculation section follows the same steps as in the above example. The required function values are summed in S starting at line 315 where S is set to half the first function value. The loop in lines 325 to 340 adds the other function values to S except the final value, which has to be divided by 2, and is dealt with separately in line 345. The final step of multiplying by h to find the area occurs in line 350.

Program 13.5 – To find the area under the graph of a function by the trapezium rule

```
 90 REM**PROGRAM 13.5**
100 REM**TRAPEZIUM RULE**
110 DEF FNF(X) = 1/(x+2)↑2
200 REM**INPUT RANGE AND NO. OF STRIPS**
210 PRINT "WHAT RANGE : X-LOW, X-HIGH";
215 INPUT XL, XH
220 PRINT "HOW MANY STRIPS";
225 INPUT N
300 REM**CALCULATION SECTION**
310 H = (XH - XL)/N
315 S = FNF(XL)/2
320 X = XL
325 FOR J = 1 TO N - 1
330   X = X + H
335   S = S + FNF(X)
340 NEXT J
345 S = S + FNF(XH)/2
350 A = S*H
400 REM**OUTPUT**
410 PRINT "AREA IS"; A
```

13.6 NUMERICAL INTEGRATION BY SIMPSON'S RULE

The essence of numerical integration is that it uses a finite number of function values that can easily be calculated to obtain an estimate of the area under a continuous graph. Clearly it would be better to use a curved boundary at the top of each strip provided that we can somehow choose a curve that more closely follows the true graph. By taking note of function values other than just those at either end of the strip we should be able to do this. Simpson's rule uses three function values at a time and thus the area of two adjacent strips are estimated together.

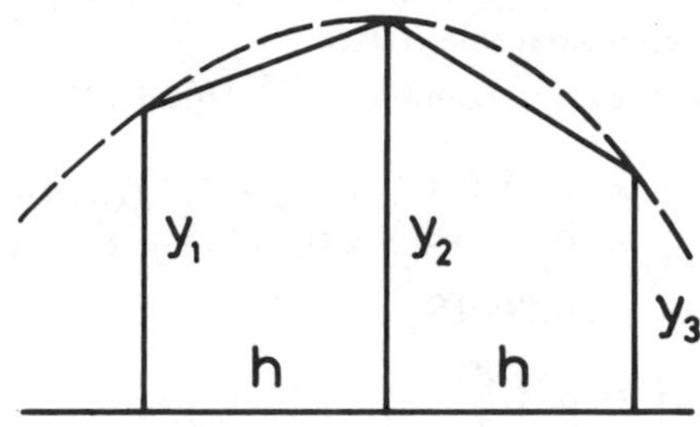

In the diagram the straight lines would be used by the trapezoidal rule but something like the dotted curve would be better. Simpson's rule chooses the quadratic through the three points for the dotted curve and it can be shown that the area under this quadratic is given by

$$(y_1 + 4y_2 + y_3) \times \frac{h}{3} \ . \tag{13.8}$$

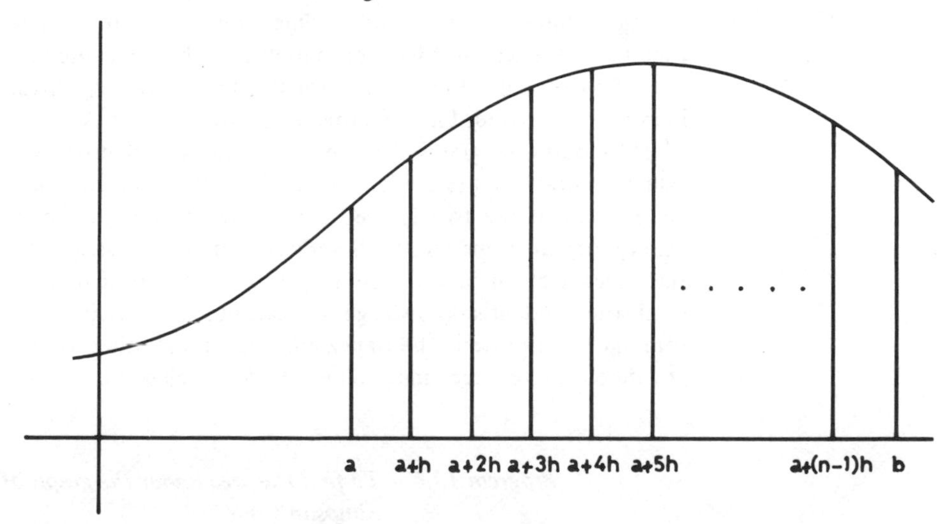

The required area is divided into an even number of strips and formula (13.8) applied to each pair in turn. The approximation to the total area is thus given by

$$\begin{aligned} A &= (f(a) + 4f(a+h) + f(a+2h) + f(a+2h) + 4f(a+3h) \\ &\quad + f(a+4h) + \ldots + f(a+(n-2)h) + 4f(a+(n-1)h) \\ &\quad + f(a+nh)) \times \frac{h}{3} \\ &= (f(a) + 4f(a+h) + 2f(a+2h) + 4f(a+3h) + 2f(a+4h) + \ldots \\ &\quad + 2f(a+(n-2)h) + 4f(a+(n-1)h) + f(a+nh)) \times \frac{h}{3} \end{aligned} \tag{13.9}$$

and this is Simpson's rule. The true curve has been replaced with a series of pieces of quadratic curves each one joining three successive points of the graph. It can be shown that the error when using Simpson's rule is very much smaller than when the trapezoidal rule is used.

For the previous example of $\int_1^3 [1/(x+2)^2]\,dx$ if we use Simpson's rule then

$$\begin{aligned} S &= f(1) + 4 \times f(1.5) + 2 \times f(2) + 4 \times f(2.5) + f(3) \\ &= 0.11111 + 4 \times 0.08163 + 2 \times 0.0625 + 4 \times 0.04938 + 0.04 \\ &= 0.80015 \\ A &= S \times \frac{h}{3} \\ &= 0.80015 \times 0.5 \div 3 \\ &= 0.13336 \end{aligned}$$

which is a much better result than that obtained with the trapezium rule.

Program 13.6 is a simple implementation of Simpson's rule. The structure of the program is similar to Program 13.5. S is set to the first function value in line 315. The loop in lines 325 to 350 deals with two function values at a time, adding 4 times the first and 2 times the second to S as required. The final function value will also have been multiplied by 2 but the formula only requires one of these. Line 355 corrects for this by subtracting this last value after the loop has terminated. Line 360 then calculates the area A.

Computer programs for the trapezium and Simpson's rules are not only valuable tools for the student of mathematics, but also the development and comparison of the two is itself a valuable experience. They can be tried on a variety of functions and with various numbers of strips. Although Simpson's rule gives a much better accuracy, it is our belief that it should not be used until the student's knowledge of calculus is sufficiently developed for the formula to be proved. The trapezium rule, on the other hand, can be understood at a much earlier stage, and can be included in elementary courses.

Program 13.6 – To find the area under the graph of a function by Simpson's rule

```
 90 REM ** PROGRAM 13.6 **
100 REM**SIMPSON'S RULE**
110 DEF FNF(X) = 1/(X + 2)↑2
200 REM**INPUT RANGE AND NO. OF STRIPS**
210 PRINT "WHAT RANGE"
211 PRINT "X-LOW, X-HIGH";
215 INPUT XL, XH
220 PRINT "NUMBER OF STRIPS MUST BE EVEN"
221 PRINT "HOW MANY STRIPS";
225 INPUT N
300 REM**CALCULATION SECTION**
310 H = (XH - XL)/N
```

```
315  S = FNF(XL)
320  X = XL
325  FOR J = 1 TO N/2
330     X = X + H
335     S = S + 4*FNF(X)
340     X = X + H
345     S = S + 2*FNF(X)
350  NEXT J
355  S = S - FNF(X)
360  A = S*H/3
400  REM**OUTPUT SECTION**
410  PRINT "AREA IS"; A
```

14

Approximating functions

14.1 DIFFERENCE TABLES

Many historical references are made to Charles Babbage's role as the founder of computing with, first, the mechanical Difference Engine and, subsequently, the planned, but never completed, Analytic Engine. The Difference Engine was designed as a machine for the efficient tabulation of polynomial functions such as those used for life expectancy tables by insurance companies, but the underlying mathematical principle had been well known long before Babbage's time.

Take any polynomial function, for example $f(x) = 4x^3 - 6x^2 - 5x + 3$ and tabulate it for a set of values of x that are at equal intervals. Suppose we use values of x from -1 to 3 in steps of 0.5 – which we shall notate as $x = -1(0.5)3$. An efficient way of evaluating $f(x)$ by hand or with a calculator is to use 'nested multiplication', that is $f(x) = ((4x - 6)x - 5)x + 3$ which reduces the arithmetic to 3 multiplications and 3 additions/subtractions per value of x. We subtract each value of the function from its successor to find the first differences Δx. These are then treated in the same way to find the second differences $\Delta^2 x$, and so on. In this case we find that the third set of differences $\Delta^3 x$ are constant. In general the result is that a polynomial of nth degree when tabulated at equal intervals yields constant nth differences $\Delta^n x$. From this result we can investigate further how the values of this constant difference are related to the form of the original polynomial and the range of values used for x. For an nth degree polynomial whose leading term is a_0x^n and which is tabulated in steps h of the independent variable x then the constant value of the nth differences is $\Delta^n x = a_0 h^n n!$ For our example above $a_0 = 4, h = 0.5$ and $n = 3$ so that

$$\Delta^n x = a_0 h^n n! = 4 \times (0.5)^3 \times 3! = 4 \times 0.125 \times 6 = 3$$

which agrees with the result in Table 14.2.

Table 14.1

x	−1	−0.5	0	0.5	1	1.5	2	2.5	3
f	−2	3.5	3	−0.5	−4	−4.5	1	15.5	42

Table 14.2

x	−1		−0.5		0		0.5		1		1.5		2		2.5		3
f	−2		3.5		3		−0.5		−4		−4.5		1		15.5		42
Δx		5.5		−0.5		−3.5		−3.5		−0.5		5.5		14.5		26.5	
$\Delta^2 x$			−6		−3		0		3		6		9		12		
$\Delta^3 x$				3		3		3		3		3		3			
$\Delta^4 x$					0		0		0		0		0				

However, the power of the general result is that once enough of the table has been constructed in order to find a value of the nth difference then all subsequent values of the polynomial can be found by continuing the pattern in the $\Delta^n x$, $\Delta^{n-1}x$, ..., Δx and finally the f rows. This process requires just n additions per step compared with the $n - 1$ multiplications and n additions for nested multiplication and $n(n + 1)/2$ multiplications and n additions for the conventional means of evaluation.

In the present example assuming that $\Delta^3 x$ stays constant at +3 then the next value of $\Delta^2 x = +12 + (+3) = +15$, the next value of $\Delta x = +26.5 + (+15) = +41.5$ and hence the next value of $f = +42 + (+41.5) = 83.5$ which agrees with $f(3.5)$ computed by normal means.

This, then, was the basis of the design of Babbage's Difference Engine which utilised a system of rows of cogs – one for each row of differences. Each turn of a row of cogs enmeshed with the next to add on a quantity. By setting the first row of cogs to represent the constant value of the nth differences the machine modelled the addition of differences to produce the function values.

In mathematical investigations we may frequently collect data that, from the formation of a difference table, appear to conform to a polynomial pattern. Using difference tables we can determine the form of the polynomial function which would generate identical values. Suppose we have a function that produces the difference table shown in Table 14.3.

Table 14.3

x	0		1		2		3		4		5
f	3		−7		−25		−39		−37		−7
Δx		−10		−18		−14		2		30	
$\Delta^2 x$			−8		4		16		28		
$\Delta^3 x$				12		12		12			

Assuming that that $\Delta^3 x$ is constant we deduce that $f(x)$ is a cubic of the form $f(x) = a_0 x^3 + a_1 x^2 + a_2 x + a_3$. By using four consecutive values of x we can produce 4 equations in 4 unknowns:

$$\begin{aligned} a_3 &= 3 \\ a_0 + a_1 + a_2 + a_3 &= -7 \\ 8a_0 + 4a_1 + 2a_2 + a_3 &= -25 \\ 27a_0 + 9a_1 + 3a_2 + a_3 &= -39 \end{aligned}$$

In general, then, we would have to solve $n + 1$ equations in $n + 1$ unknowns to find the coefficients of an nth degree polynomial $f(x)$. Knowing that $\Delta^n x = a_0 n! h^n$ helps to find one of the coefficients directly but the others still need to be determined by considerable computation. Fortunately there is a much more convenient way of arriving at the form of the function.

From the above discussion it is clear that we can condense a difference table down dramatically to a very portable form. For example from a forward pointing diagonal:

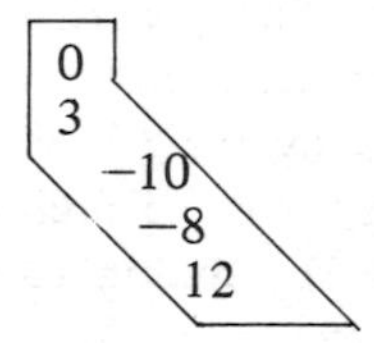

the whole table may be generated – forwards by addition and backwards by subtraction. The form of this portable condensation of the difference table gives rise our nomenclature of a 'nought-sock'. If we treat this as a vector we could decompose it as a linear combination of the set of elementary (or basis) vectors

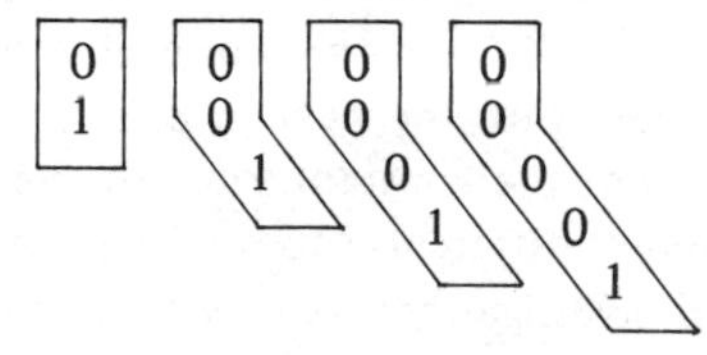

etc. Now

0
1

is the portable form of the table:

x	0	1	2	3	4	...
f_0	1	1	1	1	1	...

which is that of the constant function $f_0(x) = 1$. Similarly

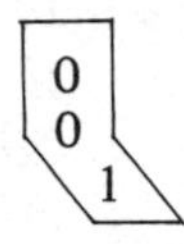

represents the table:

x	0	1	2	3	4	5	...
f_1	0	1	2	3	4	5	...
Δx	1	1	1	1	1	...	

where $f_1(x) = x$ is a linear function.

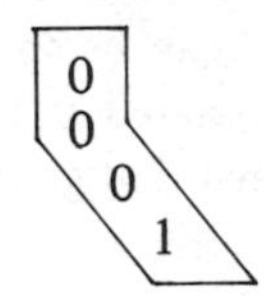

produces the table:

x	0	1	2	3	4	...
f_2	0	0	1	3	6	...
Δx	0	1	2	3	...	
$\Delta^2 x$	1	1	1	...		

which is that of a quadratic function which has roots at $x = 0$ and at $x = 1$. Hence $f_2(x) = kx(x-1)$, but since $f_2(2) = 1 = 2k$ we have that $f_2(x) = x(x-1)/2$. The 'sock'

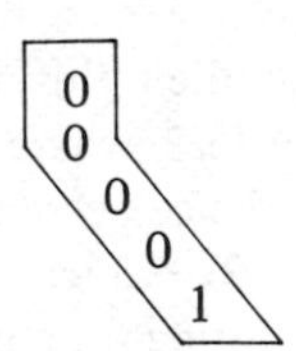

gives the table:

x	0	1	2	3	4	5	...;
f_3	0	0	0	1	4	10	...;
Δx	0	0	1	3	6	...	
$\Delta^2 x$	0	1	2	3	...		
$\Delta^3 x$	1	1	1	...			

which is that of a cubic function with roots at $x=0, x=1$, and $x=2$. Thus $f_3(x) = kx(x-1)(x-2)$ and $f_3(3) = k.3.2.1 = 1$, that is $k = 1/3!$ and so $f_3(x) = [x(x-1)(x-2)]/3!$ and the form of the other basis polynomials should now be clear. This set of polynomials is known as 'factorial polynomials'. Decomposing

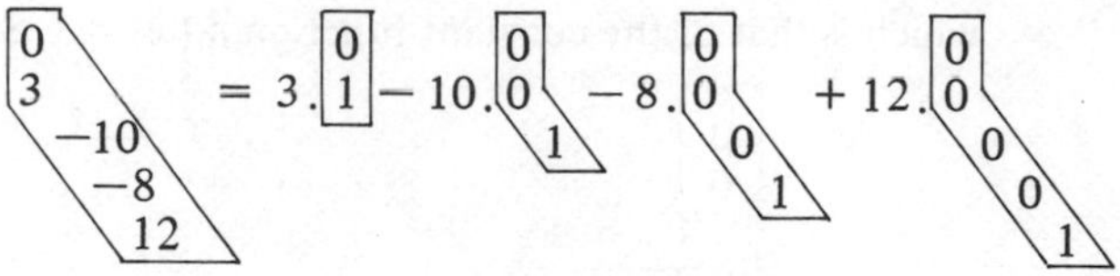

$$\begin{matrix}0\\3\\&-10\\&&-8\\&&&12\end{matrix} = 3.\begin{matrix}0\\1\end{matrix} - 10.\begin{matrix}0\\0\\&1\end{matrix} - 8.\begin{matrix}0\\0\\&0\\&&1\end{matrix} + 12.\begin{matrix}0\\0\\&0\\&&0\\&&&1\end{matrix}$$

that is,

$$\begin{aligned} f(x) &= 3\times 1 - 10\times x - 8\times\frac{x(x-1)}{2} + 12\times\frac{x(x-1)(x-2)}{3!} \\ &= 3 - 10x - 4x(x-1) + 2x(x-1)(x-2) \\ &= 3 - 2x - 10x^2 + 2x^3 \end{aligned}$$

Here we have used a set of basis polynomials, the factorial polynomials, as a suitable basis for difference tables. They can be applied just as well if the values of the independent variable x are not tabulated at unit steps nor include $x = 0$ as a data value. Consider finding a polynomial that generates the following table:

x	−3		−1		1		3		5
f	2		−2		0		8		22
Δx		−4		2		8		14	
$\Delta^2 x$			6		6		6		

Assuming $\Delta^2 x$ is constant at 6 then we can decompose the 'nought-sock':

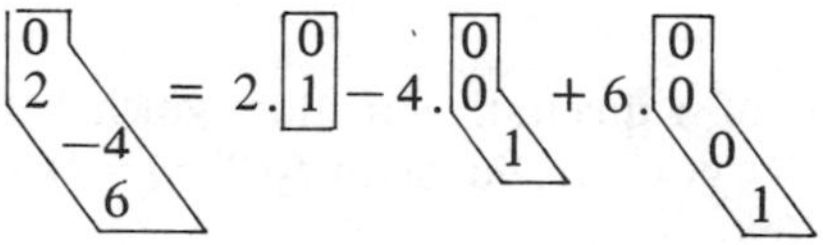

$$\begin{matrix}0\\2\\&-4\\&&6\end{matrix} = 2.\begin{matrix}0\\1\end{matrix} - 4.\begin{matrix}0\\0\\&1\end{matrix} + 6.\begin{matrix}0\\0\\&0\\&&1\end{matrix}$$

Setting z as the independent variable then this would be generated by the quadratic function $2 - 4z + 6 \times z(z-1)/2$ where $z = 0, 1, 2, 3, \ldots$ corresponds to $x = -3, -1, 1, 3, \ldots$. The correspondence between z and x is thus linear since x is in arithmetic progression, thus $x = 2z - 3$ which can be inverted to give $z = (x+3)/2$. Hence $f(x)$ is given by substituting for z in terms of x:

$$\begin{aligned} f(x) &= 2 - 4\times\frac{x+3}{2} + 6\times\frac{\frac{(x+3)}{2}\times\left(\frac{x+3}{2}-1\right)}{2} \\ &= 2 - 2(x+3) + 3\,\frac{x+3}{2}\,\frac{x+1}{2} \\ &= \frac{1}{4}(3x^2 + 4x - 7)\ . \end{aligned}$$

This property of differences of polynomials is also used to interpolate intermediate values in a table of values of a function and is often referred to as Gregory–Newton forward difference interpolation . However, the original use by Babbage as an efficient means of evaluating a polynomial is as important today as it was then. Many computer applications, such as curve fitting, require the evaluation of successive points on a polynomial and the efficiency of this process can be greatly improved by the successive addition of differences.

14.2 CHEBYCHEV POLYNOMIALS

In Chapter 11 we consider examples of series approximations to common functions such as $\sin x$, $\cos x$, e^x, $\ln(1+x)$ etc. Many programming languages include a set of standard functions such as SIN, COS, EXP, LOG etc. and, obviously, the compiler or interpreter for the language must contain algorithms for evaluating these functions. A more obvious example is that many cheap calculators have such functions available at the touch of a button. Obviously some functions do take a calculator or computer longer to evaluate than others. In the search to make the evaluation of functions more efficient the standard Taylor series representations are usually altered somewhat. Consider computing $\sin x$: the normal series representation is $\sin x = x - x^3/3! + x^5/5! - \ldots$ where x is in radians. When x is small only a few terms of this series will be needed to give a reasonable degree of accuracy, but many more terms will be needed to produce the same accuracy for large value of x. Also, because of the different numbers of terms required, any algorithm to produce $\sin x$ to a given accuracy will have to contain a test which will be applied after each term of the series has been generated which will, in itself, slow the process up. It would be nice if we could smooth out the approximation errors in Taylor series approximations so that a series with a fixed number of terms will give the same accuracy over a range of values of x.

As an example consider evaluating $\sin 0.1$ to 3 decimal places. To do this we can cut off the series just before a term $x^{2r+1}/(2r+1)!$ which has an absolute value less than half a unit of the third decimal place, that is

$$\frac{0.1^{2r+1}}{(2r+1)!} < 0.0005$$

$$(0.1)^{2r+1} < 0.0005 \times (2r+1)!$$

$$2000 < 10^{2r+1} \times (2r+1)!$$

and so $r = 1$ is the smallest suitable value for r giving $\sin x = x$ as an approximation accurate to 3D for $x = 0.1$. Consider, now, $x = 10$. From the previous argument we have

$$\frac{10^{2r+1}}{(2r+1)!} < 0.0005$$

$$2000 \times 10^{2r+1} < (2r+1)!$$

and we find that $r = 16$ is the smallest suitable value, which means we have to go as far as the term $-x^{31}/31!$ to achieve the desired accuracy.

The basis of the smoothing technique is the set of Chebychev polynomials. The first few of these are:

$$T_0(x) = 1$$
$$T_1(x) = x$$
$$T_2(x) = 2x^2 - 1$$

and they are defined, generally, by the rather strange-looking formula:

$$T_n(x) = \cos(n \cos^{-1} x)$$

As the function involves the inverse cosine of an angle it is only defined for $-1 \leqslant x \leqslant 1$. Since it is itself the cosine of angle it can only take values between -1 and 1, that is $-1 \leqslant T_n(x) \leqslant 1$. Fortunately there is a useful recurrence relationship which makes the generation of these Chebychev polynomials much easier.

Since

$$\cos(r+1)A + \cos(r-1)A = 2\cos A \cos rA$$

we have

$$T_{r+1}(x) + T_{r-1}(x) = 2xT_r(x) \qquad \text{where } x = \cos A$$

thus

$$T_{r+1}(x) = 2xT_r(x) - T_{r-1}(x)$$

which gives

$$T_3(x) = 2xT_2(x) - T_1(x) = 2x(2x^2-1) - x = 4x^3 - 3x$$
$$T_4(x) = 2x(4x^3-3x) - (2x^2-1) = 8x^4 - 8x^2 + 1 \qquad \text{etc.}$$

It is very interesting to plot successive curves $y = T_r(x)$ on the same axes for $-1 \leqslant x \leqslant 1, -1 \leqslant y \leqslant 1$.

These polynomials form another suitable set of basis polynomials from which *any* polynomial can be expressed as a linear combination of the $T_r(x)$. In order to facilitate the change of representation it is convenient to be able to express the powers of x as combinations of the $T_r(x)$, as in Table 14.4.
As an example of the 'error-smoothing' process consider the truncated approximation to $\sin x$ given by $f(x) = x - x^3/6$. This has a truncation error of approximately $x^5/5!$. For the interval $-1 \leqslant x \leqslant 1$ the maximum value of this term is $1/120 \approx 0.0083$ and hence the approximation cannot be relied on to give an accuracy of two decimal places throughout the range. Now if we include the truncated term $x^5/5!$ to give $g(x) = x - x^3/6 + x^5/120$ then we can express $g(x)$ in terms of the Chebychev polynomials $T_r(x)$ by substituting values for x, x^3 and x^5 from Table 14.4, that is

$$g(x) = T_1 - \frac{1}{24}(T_3 + 3T_1) + \frac{1}{1920}(T_5 + 5T_3 + 10T_1)$$
$$= \frac{169}{192}T_1 - \frac{5}{128}T_3 + \frac{1}{1920}T_5 \ .$$

Table 14.4

1	T_0
x	T_1
x^2	$\frac{1}{2}(T_2 + T_0)$
x^3	$\frac{1}{4}(T_3 + 3T_1)$
x^4	$\frac{1}{8}(T_4 + 4T_2 + 3T_0)$
x^5	$\frac{1}{16}(T_5 + 5T_3 + 10T_1)$
x^6	$\frac{1}{32}(T_6 + 6T_4 + 15T_2 + 10T_0)$

Since T_r is a consine of an angle the maximum absolute value of T_r is 1. If we now truncate $g(x)$ to $h(x) = (169/192)T_1 - (5/128)T_3$ we have a truncation error whose principal term is $T_5/1920$ and whose maximum value is thus $1/1920 \approx 0.0005$. Thus the cubic approximation to $\sin x$ produced by the use of Chebychev polynomials $h(x) = 0.9974x - 0.1562x^3$ has a maximum truncation error of approximately 0.0005 in the range $[-1, 1]$ compared with the corresponding Taylor series approximation $g(x) = x - 0.1667x^3$ which has a maximum truncation error of about 0.008 in the same range.

This process is sometimes known as the 'economization' of polynomials. Applying it to fifth degree approximations to $\sin x$ we have:

Chebychev:

$$h(x) = 0.999979x - 0.166492x^3 + 0.007984x^5$$

max. error: 0.000003

Taylor:

$$g(x) = 1.000000x - 0.166667x^3 + 0.008333x^5$$

max. error: 0.000192

At first sight the restriction $-1 \leqslant x \leqslant 1$ might seem rather harsh. However, with efficient formulae for both $\sin x$ and $\cos x$ in the range $-1 \leqslant x \leqslant 1$ we can evaluate the sine or cosine of any angle by reducing it to lie in the required range. Consider $\sin 1000°$ – first reduce $1000°$ by full circles until it lies in the range $(-180°, 180°]$, that is $\sin 1000° = \sin(-80°)$. This is still not in the required range since 1 radian is roughly $57°$ – but $\sin(-80°) = -\sin 80° = -\cos 10°$. Hence $\sin 1000°$ can be computed by putting $x = \pi/18$ in the economised version of the cosine series.

14.3 B-SPLINES

When trying to find a mathematical formulation to approximate a complex curve it may be useful to consider the curve as consisting of a number of separate curves and to find approximations to each piece. A mechanical analogy to this process is given by a draughtsman's instrument called a *spline*. This consists of a long piece of springy steel along which a set of heavy metal weights, called *ducks*, may be moved. The ducks are positioned at appropriate points on a drawing board and the springy steel spline then takes up the shape of a smooth curve. Because of the nature of steel and the constraints imposed by contact at the ducks the overall shape of the curve cannot be described by a single simple polynomial function, but each *span* between ducks does have a simple polynomial representation. This *piecewise* representation of the spline must, of course, satisfy properties of continuity and continuity of derivatives at the ducks (or *knots* as they are also known). This approach to the piecewise description of curves has been frequently applied in computer graphics over the past few years. As with factorial polynomials and Chebychev polynomials it is convenient to define a suitable set of *basis* polynomials to use as the building blocks for splines and it is this set which is known as the B-splines.

The simplest B-spline $B_0(x)$ has a value of 1 if $-\frac{1}{2} \leqslant x \leqslant \frac{1}{2}$ and 0 everywhere else. This piecewise definition can be written:

$$B_0(x) = \begin{cases} 1 & \text{if } -\frac{1}{2} \leqslant x \leqslant \frac{1}{2} \\ 0 & \text{otherwise} \end{cases}$$

and its graph has the form shown in Fig. 14.1.

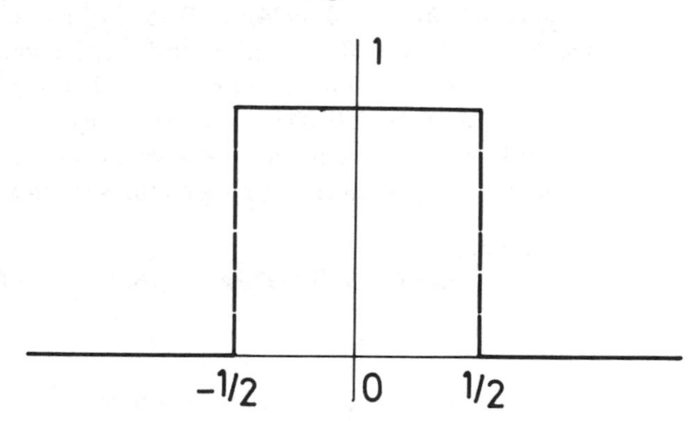

Fig. 14.1

The linear B-spline $B_1(x)$ has a triangular shape with values of $B_1(0) = 1$ and $B_1(-1) = B_1(1) = 0$. Its definition is:

$$B_1(x) = \begin{cases} 1-x & \text{if } 0 \leqslant x \leqslant 1 \\ 1+x & \text{if } -1 \leqslant x \leqslant 0 \\ 0 & \text{otherwise} \end{cases}$$

Its graph is shown in Fig. 14.2 and the area under it is 1.

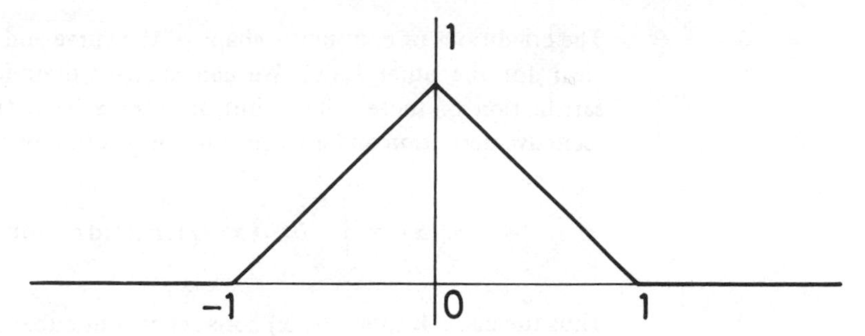

Fig.14.2

The quadratic B-spline $B_2(x)$ consists of three quadratic arcs in the intervals $[-1\frac{1}{2},-\frac{1}{2}]$, $[-\frac{1}{2},\frac{1}{2}]$, $[\frac{1}{2},1\frac{1}{2}]$ such that

(i) the area under the curve is 1;
(ii) the function is continuous at $x=-\frac{1}{2}$ and $x=\frac{1}{2}$;
(iii) the arcs have gradient zero at $x=1\frac{1}{2}$, $x=0$ and $x=-1\frac{1}{2}$ and
(iv) the gradients of the tangents of the arcs which meet at $x=\frac{1}{2}$ are equal there and similarly at $x=-\frac{1}{2}$.

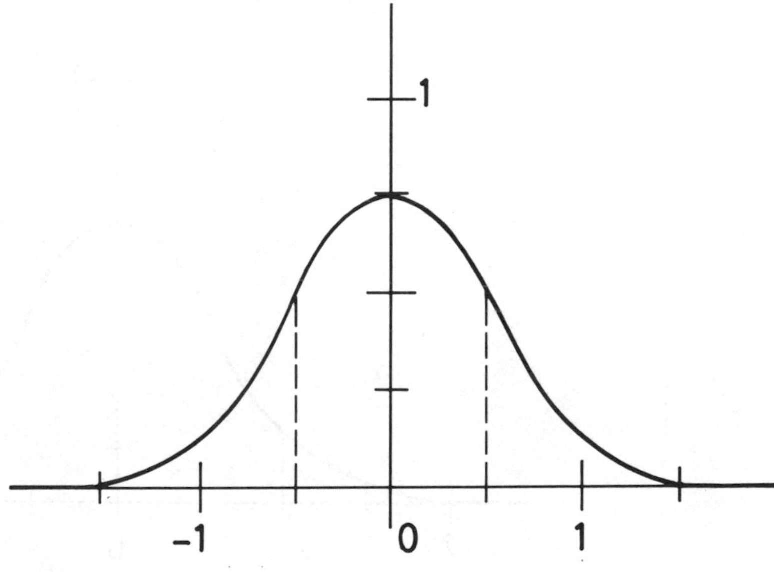

Fig. 14.3

This information is enough to specify the quadratic arcs completely and thus we can derive its piecewise form as:

$$B_2(x) = \begin{cases} \frac{1}{2}(x-1\frac{1}{2})^2 & \text{if} \quad \frac{1}{2} \leqslant x \leqslant 1\frac{1}{2} \\ \frac{3}{4}-x^2 & \text{if} \quad -\frac{1}{2} \leqslant x \leqslant \frac{1}{2} \\ \frac{1}{2}(x+1\frac{1}{2})^2 & \text{if} \quad -1\frac{1}{2} \leqslant x \leqslant -\frac{1}{2} \\ 0 & \text{otherwise} \end{cases}$$

The conditions of continuity, shape of the curve and area under it should now be clear for the other $B_r(x)$. We can identify their form from the simultaneous satisfaction of these criteria but an alternative definition is available. This is a recursive definition and involves the convolution of functions:

$$B_n(x) = \int_{-\infty}^{\infty} B_{n-1}(x-t).B_0(t)\mathrm{d}t \quad \text{for} \quad n = 1, 2, 3, \ldots.$$

Thus the cubic B-spline $B_3(x)$ consists of four cubic arcs and has as definition:

$$B_3(x) = \begin{cases} \frac{1}{6}(2-x)^3 & \text{if} \quad 1 \leqslant x \leqslant 2 \\ \frac{2}{3} - x^2 + \frac{1}{2}x^3 & \text{if} \quad 0 \leqslant x \leqslant 1 \\ \frac{2}{3} - x^2 - \frac{1}{2}x^3 & \text{if} \quad -1 \leqslant x \leqslant 0 \\ \frac{1}{6}(2+x)^3 & \text{if} \quad -2 \leqslant x \leqslant -1 \\ 0 & \text{otherwise} \end{cases}$$

and its graph is shown in Fig. 14.4.

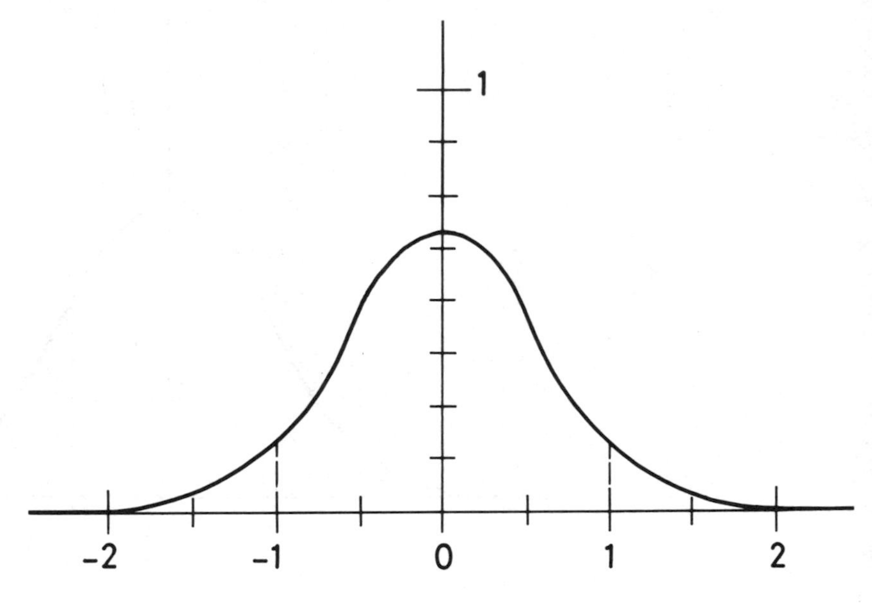

Fig. 14.4

It appears that the shape of these graphs is approaching that of a normal distribution. This is indeed the case: B_0 is the probability density function (p.d.f.) of a uniform distribution, B_1 is the p.d.f. of the sums of pairs of samples from the uniform distribution, B_2 is the p.d.f. of the sums of three samples etc., and thus the Central Limit Theorem shows that B_n tends to a normal distribution with mean 0 and variance $(n + 1)/12$. In section 9.7.1 we used $B_{11}(x)$ as an approximation to the standard normal distribution.

The B-splines can be used to generate weighting functions for use in taking weighted averages of a number of quantities. Consider the quadratic B-spline $B_2(x)$ and look at the y co-ordinates of points in 'corresponding' positions in each of the three arcs. Suppose we consider points a quarter the way along each arc then, for the first arc, we have an x value a quarter of the way from $x = \frac{1}{2}$ to $x = 1\frac{1}{2}$, that is, $x = \frac{3}{4}$ – substituting this in B_2 gives a value of 9/32 as the y co-ordinate. Similarly in the second arc a point a quarter of the way from $x = -\frac{1}{2}$ to $x = \frac{1}{2}$ is at $x = -\frac{1}{4}$ for which $y = 11/16$ and in the third arc we have $x = -1\frac{1}{4}$ and $y = 1/32$. The three quantities: 1/32, 11/16 and 9/32 have a total of 1 and so, indeed, can be used as weights. Using a parameter t with a range of values from 0 and 1 in place of the particular value of $\frac{1}{4}$ gives the functions $f_1(t) = \frac{1}{2}t^2$, $f_2(t) = \frac{1}{2} + t - t^2$ and $f_3(t) = \frac{1}{2}(1-t)^2$ which are all non-negative for $0 \leqslant t \leqslant 1$ and which total 1 and so can form weights. For the cubic B-spline $B_3(x)$ the corresponding weighting functions are:

$$f_1(t) = t^3/6 \; , \quad f_2(t) = \frac{(1+t)^3 - 4t^3}{6} \; ,$$

$$f_3(t) = \frac{(2-t)^3 - 4(1-t)^3}{6} \; , \quad \text{and} \quad f_4(t) = \frac{(1-t)^3}{6} \; .$$

These weighting functions can be used to generate curved arcs defined by a set of points. Suppose the three points P_1, P_2 and P_3 have position vectors $\mathbf{p}_1$, $\mathbf{p}_2$ and $\mathbf{p}_3$ then we can define an arc generated by these points by giving the position vector $\mathbf{p}$ of a point P which lies on it as the weighted average of the position vectors of P_1, P_2 and P_3. If we use the weighting functions derived from the quadratic B-spline then we have $\mathbf{p}$ as a function of t:

$$\mathbf{p}(t) = f_1(t) . \mathbf{p}_1 + f_2(t) . \mathbf{p}_2 + f_3(t) . \mathbf{p}_3 \qquad 0 \leqslant t \leqslant 1$$

When

$t = 0$ $\quad \mathbf{p}(0) = \frac{1}{2}\mathbf{p}_2 + \frac{1}{2}\mathbf{p}_3$ which is the mid-point of P_2P_3

$t = 1$ $\quad \mathbf{p}(1) = \frac{1}{2}\mathbf{p}_1 + \frac{1}{2}\mathbf{p}_2$ the mid-point of P_1P_2

$t = \frac{1}{2}$ $\quad \mathbf{p}(\frac{1}{2}) = \frac{1}{8}\mathbf{p}_1 + \frac{3}{4}\mathbf{p}_2 + \frac{1}{8}\mathbf{p}_3$ which is one-quarter the way from P_2 along the median P_2P_4 of the triangle $P_1P_2P_3$.

We can differentiate the expression for $\mathbf{p}(t)$ with respect to t to find tangent vectors to the curved arc. Thus

$$\mathbf{p}'(t) = t\mathbf{p}_1 + (1-2t)\mathbf{p}_2 - (1-t)\mathbf{p}_3$$

So putting $t = 0, t = 1$ and $t = \frac{1}{2}$ we have

$\mathbf{p}'(0) = \mathbf{p}_2 - \mathbf{p}_3$ that is parallel to P_3P_2

$\mathbf{p}'(1) = \mathbf{p}_1 - \mathbf{p}_2$ parallel to P_2P_1

$\mathbf{p}'(\frac{1}{2}) = \frac{1}{2}(\mathbf{p}_1 - \mathbf{p}_3)$ parallel to P_3P_1

and thus the curve behaves as shown in Fig. 14.5. The reader might like to explore the behaviour of an arc generated by 4 points using the cubic B-spline weighting functions as this is the most commonly used spline representation.

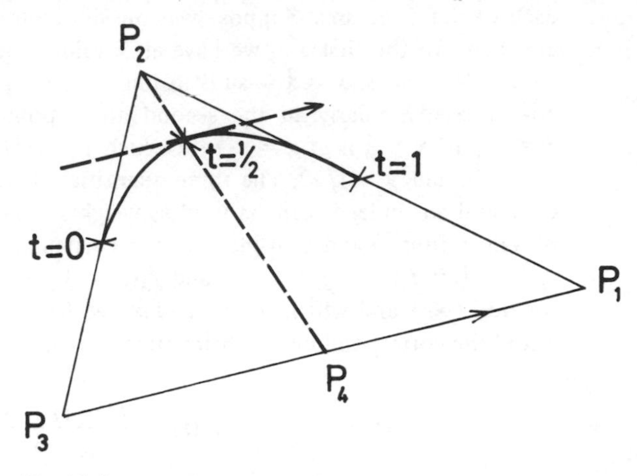

Fig. 14.5

Now that we have defined an arc generated by r points using $(r-1)$th degree B-spline weights we can define a curve generated by n points as a set of arcs formed by taking different groupings of r points. As an example consider again using quadratic B-spline arcs generated by 3 points at a time to fit a curve to the n points $P_1, P_2, P_3, \ldots, P_n$. An arc can be generated from $P_1P_2P_3$ another from $P_2P_3P_4$ and so on up to $P_{n-2}P_{n-1}P_n$. These $n-2$ arcs are continuous and have common tangents at their joins but their second derivatives are, in general, discontinuous at the joins. Each of the interior points $P_3, P_4, \ldots, P_{n-2}$ contributes to *three* of the quadratic arcs, whereas P_2 and P_{n-1} contribute to two arcs and P_1 and P_n contribute to one arc each. As with the Bezier curves of Chapter 6 the curve is controlled by the set of points but it does not, in general, pass through them. The key difference between Bezier and B-spline curves is that when a point such as P_4 is moved the whole of the Bezier curve is changed whereas just three of the arcs in this quadratic B-spline representation are changed – this property of B-spline curves is called *local control*. It is important to note that the points do not have all to be distinct and some choices of the set of points will give twisted curves, cusps or closed curves. In order to make each of the points of the set contribute to the same number of arcs it is customary to extend the set of points to (i) $P_1, P_1, P_1, P_2, P_3, \ldots, P_{n-2}, P_{n-1}, P_n, P_n, P_n$ for an open quadratic curve or $P_1, P_1, P_1, P_1, P_2, P_3, \ldots, P_{n-2}, P_{n-1}, P_n, P_n, P_n, P_n$ for an open cubic curve curve etc. and (ii) $P_1, P_2, P_3, \ldots, P_{n-1}, P_n, P_1, P_2$ for a closed quadratic curve or $P_1, P_2, P_3, P_4, \ldots, P_{n-1}, P_n, P_1, P_2, P_3$ for a closed cubic curve etc. Fig. 14.6 shows a quadratic B-spline curve generated by 4 points using (i).

A similar approach can be adopted to defining surfaces as functions of two parameters rather than curves as a function of one parameter. The little pieces of surface corresponding to arcs are called *patches* and many systems for computer aided design of surfaces (such as car body panels) now use bi-cubic B-splines as the representation for patches.

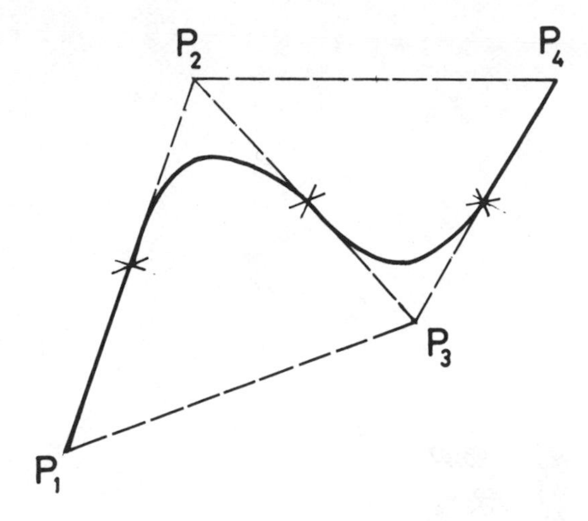

Fig. 14.6

An alternative definition of the B-splines can be given by inventing the function $(x)_+^n$ which has a piecewise definition:

$$(x)_+^n = \begin{cases} x^n & \text{if } x > 0 \\ 0 & \text{if } x \leqslant 0 \end{cases}$$

Thus the cubic version $(x)_+^3$ has the form shown in Fig. 14.7. The cubic B-spline function, $B_3(t)$ can now be defined as:

$$B_3(t) = \frac{1}{6}\left[(t+2)_+^3 - 4(t+1)_+^3 + 6t_+^3 - 4(t-1)_+^3 + (t-2)_+^3\right]$$

and, from this form, it is easy to see the structure of the general B-spline.

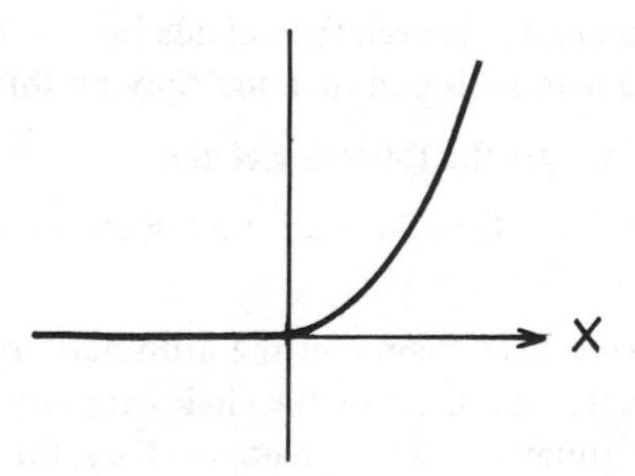

Fig. 14.7

15

Numerical investigations

15.1 NUMBER CHAINS

Playing with numbers has a fascination for many people and can also lead to some worthwhile mathematical activity. One such situation is the conjecture that if you follow a certain rule the chain of numbers produced always leads to 1, though sometimes the chain can be quite long and tortuous. The rule is in two parts:

If a number is even then divide by two.
If a number is odd then multiply by three and add one.

Starting with 7 we get the following chain:

$$7 \rightarrow 22 \rightarrow 11 \rightarrow 34 \rightarrow 17 \rightarrow 52 \rightarrow 26 \rightarrow 13 \rightarrow 40 \rightarrow 20 \rightarrow 10 \rightarrow 5 \rightarrow 16 \rightarrow$$
$$\rightarrow 8 \rightarrow 4 \rightarrow 2 \rightarrow 1$$

This is not a very long chain and the arithmetic is quite easy so no calculating device is desirable, but some of the chains are very much longer, and the investigation of the situtation can be hampered by the tedious arithmetic involved. A simple computer program can be devised to take the drudgery out of the situation. We need a test for odd or even and this can be done with the INT function.† In Program 15.1 this is done in line 130. By the use of this simple

† An interesting alternative to the use of the INT function is the use of logical AND suggested in Chapter 7. Line 130 then becomes:

```
130  IF (N AND 1) = 0 THEN GOTO 150
```

program the investigation of the situation can be taken much further without the tediousness of the calculations killing the interest. It is also possible to ask what happens if we change the rules. Do other similar rules also give chains that always lead to one?

Program 15.1 – To generate a chain by the* ÷2 *or* ×3 + 1 *rule

```
 90 REM**PROGRAM 15.1**
100 REM**/2 OR *3 + 1 CHAINS**
110 PRINT "WHAT STARTING NUMBER";
115 INPUT N
120 PRINT N;
125 IF N = 1 THEN END
130 IF INT(N/2) = N/2 THEN GOTO 150
135 N = N*3 + 1
140 GOTO 120
150 N = N/2
155 GOTO 120
```

What about other kinds of rules? Playing with the digits in a number can lead to some fascinating patterns. If we multiply the units digit of a two-digit number by a fixed multiplier and add the tens digit we get a new two digit number. Starting with 21 and using 7 as the multiplier we get the chain:

$$21 \rightarrow 9 \rightarrow 63 \rightarrow 27 \rightarrow 51 \rightarrow 12 \rightarrow 15 \rightarrow 36 \rightarrow 45 \quad \text{etc.}$$

and eventually get back to 21. To make a computer produce the chain the INT function can again be used. In Program 15.2 in lines 140 and 145 the tens and units digits are obtained. Line 150 then applies the rule using the multiplier M. Since only integers are involved integer stores may be used throughout in which case the INT function may be omitted. In lines 135 and 155 the number is turned into a string variable and this is then printed. This avoids the printing of unnecessary spaces and many numbers will appear on each line. The test in line 160 compares the latest number with the starting number and if they are different the rule is applied again. With this program different multipliers and many starting numbers may be tried in a short time.

Program 15.2 – To produce chains from the rule: multiply units digit by* M *and add tens digit

```
 90 REM**PROGRAM 15.2**
100 REM**UNITS*M + TENS CHAIN**
110 PRINT "WHAT MULTIPLIER";
115 INPUT M
120 PRINT "WHAT STARTING NUMBER";
125 INPUT S
130 N = S
135 N$ = STR$(N) : PRINT N$;"  ";
140 T = INT(N/10)
```

```
145 U = N - T*10
150 N = T + U*M
155 N$ = STR$(N) : PRINT N$;" ";
160 IF N <> S THEN GOTO 140
```

Of course the chain may not return to the starting number, but, because it will sooner or later run out of possible numbers, it must eventually return to some number that appears earlier in the chain. To test for this we need to provide for the computer to remember the whole chain in an array. In Program 15.3 the numbers are stored in array L which is dimensioned in line 105 to provide for up to 200 numbers. The numbers in the chain are counted in store C and the loop in lines 160 to 170 tests the new number N against all previous numbers in the list.

Program 15.3 – Lines to be added to Program 15.2 to compare each new number with all previous numbers in the chain

```
 90 REM**PROGRAM 15.3**
105 DIM L(200)
132 C = 1
133 L(1) = N
157 C = C + 1
158 L(C) = N
160 FOR J = 1 TO C - 1
165    IF N = L(J) THEN GOTO 190
170 NEXT J
175 GOTO 140
190 END
```

It is left to the reader to exploit the possibilities of this program and perhaps devise some alternative rules. One possible extension of the idea is to use a different number base and the program can be modified to do this. The digits will need to be typed separately but this allows for such input as a digit of say 13 in base 15.

Program 15.4 – Changes to Programs 15.2 or 15.3 to provide for numbers in any base

```
 90 REM**PROGRAM 15.4**
107 PRINT "WHAT BASE";
108 INPUT B
120 PRINT "WHAT FIRST DIGIT FOR THE"
121 PRINT "STARTING NUMBER";
122 INPUT T
123 PRINT "WHAT SECOND DIGIT FOR THE"
124 PRINT "STARTING NUMBER";
125 INPUT U
130 N = T*B + U
140 T = INT(N/B)
145 U = N - T*B
```

Although Program 15.4 is working in the base B the output is printed in base ten. The user may wish to see the numbers in the base being used in which case it is the values of T and U that should be printed in place of N.

15.2 CHAINS OF LAST DIGITS

If we start with a number and keep multiplying by the same multiplier the final digits of the resulting numbers form a chain. For example multiplying by 7 produces

$$8 \rightarrow 56 \rightarrow 392 \rightarrow 2744 \rightarrow 19208 \rightarrow 134456 \rightarrow$$

and the last digits are

$$8 \rightarrow 6 \rightarrow 2 \rightarrow 4 \rightarrow 8 \rightarrow 6 \text{ etc.}$$

The chain contains a cycle of four numbers that repeat without end. Some digits are missing from this chain so what happens to these? 0 is of course trivial and 5 just repeats on its own. The others produce the cycle

$$3 \rightarrow 1 \rightarrow 7 \rightarrow 9 \rightarrow 3 \text{ etc.}$$

To investigate other multipliers hardly requires the aid of a computer unless large multipliers are to be tried, and even then a calculator is sufficient. An extension of the investigation is to other number bases. It then becomes tedious to do the arithmetic by hand and a computer program may be used. Since we are only interested in the final digit it does not matter what base is used in the storage of the numbers, the final digit in base B being the remainder on division by B. Program 15.5 provides for the starting number to be input in base ten in line 135. It is sufficient for this to be a single digit. Line 210 does the multiplication and then the final digit in base B is found in lines 215 and 220 and stored as N. To guard against the chain repeating only part of itself the whole list of previous numbers is searched for a repeat by the loop in lines 250 to 260. With this program the situation may be explored to a greater extent, and conjectures about the patterns produced may be made, tested and perhaps even proved.

Program 15.5 – To produce a chain of final digits when multiplying fixed number and working in a chosen base

```
 90 REM**PROGRAM 15.5**
100 REM**MULTIPLICATION CHAIN**
105 DIM L(200)
110 PRINT "WHAT BASE";
115 INPUT B
120 PRINT "WHAT MULTIPLIER";
125 INPUT M
130 PRINT "WHAT STARTING NUMBER";
135 INPUT L(1)
140 C = 1 : N = L(1)
145 N$ = STR$(N)
```

```
150 PRINT N$;" ";
200 REM**CALCULATION SECTION**
210 P = N*M
215 T = INT(P/B)
220 N = P - T*B
225 C = C + 1
230 L(C) = N
235 N$ = STR$(N)
240 PRINT N$;" ";
250 FOR J = 1 TO C - 1
255    IF N = L(J) THEN GOTO 290
260 NEXT J
265 GOTO 210
290 END
```

15.3 THE DECIMAL REPRESENTATION OF FRACTIONS

The alternative ways of representing a fraction can give rise to both puzzlement and interest. That the decimal form for such a simple fraction as 1/3 goes on indefinitely can be a surprise when it is first met. That 1/7 = 0.142 857 142 857 with a block of six digits repeating raises questions about what happens to other fractions. The other sevenths reveal a pattern and perhaps a surprise. Here we have another situation that is full of mathematical pattern, but to compute the decimals by hand quickly takes some, if not all, of the interest out of it. A calculator can be a great help but it too is limited since it only gives a few figures, and the final one may be rounded up and is not reliable. On the computer, too, straight division leads to only a few digits that in all but the simplest cases are not enough to reveal the pattern. What is needed is a program to do what we do by hand and produce the digits one at a time, always working with exact arithmetic.

To reproduce the process of converting a fraction to its decimal form let us examine the example of 3/7.

```
   0.428
7)3.000
  2 8
    20
    14
     60
     56
      4  etc.
```

We first divide the 3 by 7 to get the whole number part which in this case is 0. In Program 15.6 this is done in line 210 and the resulting integer D is stored in the string F$. This string will be used to build up the decimal so in line 220 we also add a decimal point. In our example the 3 is now the remainder at that stage and in the program TN takes this value in line 215. We next attach a 0 to the 3 and think of it as 30; this multiplication by 10 occurs in line 225. The 30 is now divided by 7 to produce the next digit, 4, of the result and line 230 does this. Line 240 then attaches this digit to the end of the string F$. The next step is to find the remainder, which is 2 in our example, and this becomes the new value of TN in line 235. The whole process is then repeated unless the remainder is now 0 or we have enough decimal places. The test in line 250 causes a new digit to be produced if both TN is non-zero and the count is still less than P.

Program 15.6 – To convert a fraction to its decimal form

```
90  REM**PROGRAM 15.6**
100 REM**FRACTION TO DECIMAL**
110 PRINT "WHAT TOP NUMBER";
115 INPUT TN
120 PRINT "WHAT BOTTOM NUMBER";
125 INPUT BN
140 PRINT "HOW MANY DECIMAL PLACES";
145 INPUT P
200 REM**CALCULATION SECTION**
205 C = 0
210 D = INT(TN/BN)
215 TN = TN - D*BN
220 F$ = STR$(D) + "."
225 TN = TN*10
230 D = INT(TN/BN)
235 TN = TN - D*BN
240 F$ = F$ + STR$(D)
245 C = C + 1
250 IF TN <> 0 AND C < P THEN GOTO 225
260 PRINT F$
```

In the process of turning a fraction into a decimal it is the remainder at each stage that determines the value of the next digit. For the example of 3/7 given above, the remainders form the chain

$$3 \rightarrow 2 \rightarrow 6 \rightarrow 4 \rightarrow 5 \rightarrow 1 \rightarrow 3$$

which repeats. It will be found that other fractions with denominator seven produce the same chain but starting at a different number. An investigation of the decimals is equivalent to investigating the remainder chains and a program to generate these provides an alternative approach. The changes to Program 15.6 to do this are given as Program 15.7. At line 205 a record is kept of the starting number S in the chain. The test in line 250 allows the chain to continue until this starting number occurs again.

Program 15.7 – Changes to Program 15.6 to produce the chain of remainders when a fraction is converted to decimal form

```
90  REM**PROGRAM 15.7**
100 REM**FRACTION TO DECIMAL REMAINDER CHAIN**
```

```
205  S = TN
220  PRINT TN;
240  PRINT TN;
250  IF TN <> S THEN GOTO 225
```

Delete lines 140, 145, 245 and 260

15.4 PYTHAGOREAN TRIADS

The fact that certain sets of whole numbers such as 3, 4, 5 and 5, 12, 13 can form the sides of a right-angled triangle is very well known, and these two sets with one or two others tend to be over-used, particularly in examination questions. One is led to wonder what other sets of whole numbers there might be that satisfy Pythagoras' theorem. The search for such sets of numbers can lead to the prediction of more of these Pythagorean triads. As part of such activity it is reasonable to ask if we have found all the possible triads in a certain range. Trial and error on a calculator rapidly becomes too tedious and time consuming and the help of a computer may be enlisted. The suggestion may be made that the computer be set to search all possible numbers. Suppose we want to find all the triads using numbers less than some chosen number N, then it is reasonable to try some such program as Program 15.8. It is assumed that A is the smallest of the three numbers so values of A up to B − 1 are sufficient. Similarly values of B up to C − 1 are used, making C the largest of three numbers. The test in line 140 uses this fact and A, B and C will only be printed when they fit the theorem. It should be noted that the use of the exponent 2 in this line can lead to some triads being missed since on some computers A↑2 is not exact whereas for whole numbers within the range of possible integers A∗A is always exact.

***Program 15.8 – To find Pythagorean triads by direct search of all numbers up to* N**

```
 90  REM**PROGRAM 15.8**
100  REM**PYTHAGOREAN-TRIADS**
110  PRINT "WHAT MAXIMUM NUMBER";
115  INPUT N
120  FOR C = 3 TO N
125     FOR B = 2 TO C - 1
130        FOR A = 1 TO B - 1
140           IF A*A + B*B = C*C THEN PRINT A, B, C
145        NEXT A
150     NEXT B
155  NEXT C
```

Program 15.8 is beautifully simple but one soon finds that it does not perform very well. The major drawback is that the time taken becomes very large unless N is kept to an unreasonably small value. It is not usually possible or convenient to run the microcomputer all day on one problem, and in any case one would like the results sooner. The time taken can be reduced considerably by removing the A loop and noting that since we require $A^2 + B^2 = C^2$, once B and C have been chosen only the integer near to $\sqrt{C^2 - B^2}$ needs to be tested for A. Testing one A value instead of a range makes the program noticeably faster. It would not do to test $\sqrt{C^2 - B^2}$ to see if it is an integer since the calculation of a square root is not exact, and a slight corruption can be introduced into the result. Line 130 may be replaced with

```
130  A = INT(SQR(C*C - B*B) + 0.1)
```

and line 145 omitted. The program now takes less time but a new defect appears in the results. Each triad appears twice since we no longer have A less than B. A simple test to reject values of A greater than B would correct this, but we can save even more time by reducing the range of values of B that are tested. If A is to be less than B, then since $A^2 + B^2 = C^2$ we have $2B^2 > C^2$ or $B > C/\sqrt{2}$. In Program 15.9 the value of $\sqrt{2}$ is computed once only in line 117 and the smallest value of B that needs to be tested is obtained as BE in line 123.

Program 15.9 – Improved version of Program 15.8 for Pythagorean triads to speed the calculation

```
 90  REM**PROGRAM 15.9**
100  REM**PYTHGOREAN-TRIADS**
110  PRINT "WHAT MAXIMUM NUMBER";
115  INPUT N
117  SR = SQR(2)
120  FOR C = 4 TO N
123     BE = INT(C/SR) + 1
125     FOR B = BE TO C - 1
130        A = INT(SQR(C*C - B*B) + 0.1)
140        IF A*A + B*B = C*C THEN PRINT A, B, C
145     NEXT B
150  NEXT C
```

Although all the triads now produced involve different numbers, many have a common factor and are thus simply related to an earlier set in the list. For example the triad 3, 4, 5 will also appear as 6, 8, 10 and 9, 12, 15, etc. It may be considered desirable to omit from the printed list any triads having a factor in common. A subroutine can be added to the program to use the Euclidean algorithm to find the highest common factor of two numbers. The subroutine receives two numbers Q and R and returns to the main program their highest common factor in Q. In line 210 of Program 15.10, Q and R take the values of A and B so that on proceeding to line 215 their highest common factor is in Q. Line 215 gives R the value of C and the subroutine is called again. On return this time Q contains the highest common factor of all three and only if Q = 1 are the values of A, B and C printed.

Program 15.10 – Changes and additions to Program 15.9 to omit triads with a common factor

```
 90  REM**PROGRAM 15.10**
140  IF A*A + B*B = C*C THEN GOTO 200
```

```
190 END
200 REM**TEST FOR COMMON FACTORS**
210 Q = A : R = B : GOSUB 1000
215 R = C : GOSUB 1000
220 IF Q>1 THEN GOTO 150
225 PRINT A, B, C
230 GOTO 150
1000 REM**H.C.F. SUBROUTINE**
1005 IF Q<R THEN T = Q : Q = R : R = T
1010 P = Q
1015 Q = R
1020 R = P - INT(P/Q)*Q
1025 IF R>0 THEN GOTO 1010
1030 RETURN
```

It is an interesting exercise to compare the output of Program 15.10 with triads generated in other ways. One way is to use equations (15.1).

$$A = P^2 - Q^2 \quad B = 2PQ \quad C = P^2 + Q^2 \tag{15.1}$$

It can easily be shown algebraically that if P and Q are any whole numbers then these formulae produce a triad, but are all triads produced in this way? The output of Program 15.10 may also be organised in various ways to reveal sets of triads with a common pattern. This leads to the prediction of infinite sets of triads and to general formulae for these.

16

Problem solving

No attempt will be made to categorise different modes of problem solving nor to suggest classes of problem in which particular computing techniques may be applicable. We are concerned here with uses of a computer to increase the time and evidence available for mathematical symbolisation and speculation leading to the formation of hypotheses and, perhaps a search for proof. At times this may make use of rather 'bull at a gate' programs to produce data, in other cases, though, speculation may give rise to an improved program which leads to further speculation and so on. We shall give just a few examples in some detail to illustrate these notions. It is by no means claimed that the computer is essential, or necessarily desirable, in each case – that is for the reader to decide.

16.1 DIGIT SWAPPING

Can you find a number which is increased by 50% when its rightmost digit is removed and placed before its leftmost digit?

The first problem, then, is to make the computer model the process of shifting the last digit from the back to the front. We want 123 to yield 312, 9752 to give 2975, and 100 001 to produce 110 000. A roundabout approach available to those who have used BASIC for more than just number handling is the use of string functions, which, in their own way, are close analogies to the way we handle numbers as strings of digits.

Suppose X holds the number to be transformed then, first, we must turn X from a numeric variable into a string of characters – the function which performs this is STR$(X), for example if X = 357 then X$ = STR$(X) makes X$ hold the string of symbols 3, 5 and 7 in that order. The function LEN(X$) can be used to find how many symbols the string X consists of so put L = LEN(X$). The string functions LEFT$(X$,N) and RIGHT$(X$,N) allow the frontmost and backmost N symbols of X$ to be extracted. Thus D$ = RIGHT$(X$,1) should make D$ hold the extreme right-hand digit of X and L$ = LEFT$(X$,L−1) should put the remaining digits in L. The shifting of the last digit to the front can be achieved by Y$ = D$ + L$, which assembles the two sub-strings D$ and L$ in the new order. Finally the string of symbols Y$ must be turned back into a numeric variable to see if it is 50% bigger than X – this is performed by Y = VAL(Y$). Try this process for yourself. If you do not arrive at the expected transformation of X then arrange for Y$ to be printed out to see if you can detect where the problem lies. This process can be adapted to work correctly (if it does not do so already on your particular microcomputer) but it is much more instructive to see if we can arrive at a more mathematically satisfying procedure to achieve the same result.

Starting again with X we can see that L = INT(X/10) allows the last digit to be 'chopped off'. The last digit can then be found by multiplying L by 10 and subtracting it from X to find the remainder on dividing X by 10, that is D = X − 10*L sets D to the last digit. All that remains is to bring D in front of L. Consider the example of X = 357 in which L = 35 and D = 7, what combination of L and D yields 735? Similarly, if X = 2179 then L = 217 and D = 9, but how do we make 9217? In the first case 7 becomes 7 hundred and in the second 9 becomes 9 thousand so D must be multiplied by an appropriate power of 10. If N is the number of digits in X then Y = D*10↑(N−1) + L does the trick. Unfortunately 10↑(N − 1) is evaluated in the computer by a process of taking logarithms (base e), multiplying and finding antilogarithms (exponentiating) and this may well lead to small inaccuracies. Thus in a program it is better to use a loop to multiply D by 10 N−1 times.

We now have a numeric procedure to transform X into Y and now we can test if Y is 50% bigger than X by seeing if Y = 1.5*X or, equivalently, if 2*Y = 3*X. We are then in a position to produce a program, say, to produce all the 2-digit solutions (Program 16.1).

Program 16.1 – A first attempt at digit swapping

```
100  REM PROGRAM 16.1 – DIGIT SWAPPING VERSION 1
110  FOR X = 10 TO 99
120     L = INT(X/10)
130     D = X - 10*L
140     Y = 10*D + L
150     IF 2*Y = 3*X THEN PRINT X, Y
160  NEXT X
```

To modify the program to hunt for all 3-digit solutions just two lines need to be changed:

```
110  FOR X = 100 TO 999
140      Y = 100*D + L
```

In fact the test in line 150 makes it clear that D must be a multiple of 2 and hence that X must be an even number. Thus we can halve the number of searches made by changing 110 to count in steps of 2. Even so you may get frustrated at the lack of output from these programs. It would be more comforting if we could see something happening on the screen. We could, for example, print out all the 'near misses', for example those for which Y/X lies between 1.4 and 1.6; perhaps then we might be able to detect some sort of pattern. A suitable amendment might be:

```
150  R = Y/X : IF R>1.4 AND R<1.6 THEN PRINT X,Y,R
```

Each time we increase the number of digits in X by one we have to search through 10 times the previous number of cases. As X becomes bigger we may well reach the point beyond which an integer ceases to be held exactly in BASIC. It may be that, by leaving the computer running all night, you can find a solution to this particular problem. However, by changing the wording of the problem to read 'Can you find a number which is increased by 50% when its leftmost digit is removed and placed after its rightmost digit?' we can be fairly confident that the current approach is doomed to failure.

You may feel, then, that the computer has fulfilled in useful purpose; but just consider the algebraic relations already produced:

$$X = 10L + D$$
$$Y = L + 10^{N-1}D$$
$$2Y = 3X$$

Combining these we have:

$$2(L + 10^{N-1}D) = 3(10L + D)$$
$$2L + 2 \times 10^{N-1}D = 30L + 3D$$
$$(2 \times 10^{N-1} - 3)D = 28L \qquad (16.1)$$

where D is a single digit and L is an $N-1$ digit number. Now, for any value of N, $2 \times 10^{N-1} - 3$ is an odd number, for example 17, 197, 1997 etc. and $28L$ is an even number. Hence D must be an even digit, that is 2,4,6 or 8. Thus for each value of N we just need to test whether $(2 \times 10^{N-1} - 3)D$ is exactly divisible by 28. This allows us to produce quite a different program.

Program 16.2 – An improved attempt at digit swapping

```
100  REM PROGRAM 16.2 – DIGIT SWAPPING VERSION 2
110  M = 10
120  N = 2
130  Q = 2*M - 3
140  FOR D = 2 TO 8 STEP 2
```

```
150      P = Q*D
160      L = INT(P/28)
170      IF 28*L = P THEN PRINT L;D : STOP
180   NEXT D
190   M = 10*M
200   N = N + 1
210   GOTO 130
```

If there is a solution with a reasonable number of digits then Program 16.2 should find it in quite a short time. Remember, though, that the exact test used in line 170 may cause problems if the first (if any) solution had a large number of digits.

The divisibility argument can still be developed a stage further since $28 = 4 \times 7$. Thus the right-hand side of (16.1) is divisible by both 4 and 7 and, consequently, so must the left-hand side. Since $2 \times 10^{N-1} - 3$ is always odd then it is not divisible by either 4 or 2 and hence D must be divisible by 4. Thus we seek the smallest value of N for which $2 \times 10^{N-1} - 3$ is divisible by 7. When we have one solution we may well speculate whether there might be more and, if so, what their general form might be.

Try, now, solving the problem of left to right swapping. You may need the 'long integer' methods of Chapter 15 to help in a program or it might be possible just to use a calculator (or even long division!). Could we substitute a different value for 50%?

16.2 THE SMALLEST NUMBER WITH *T* DIVISORS

For any given number T can you find the smallest number with exactly T divisors?

Again we have to start with an obvious, but not simple, question – how can we write a program to list the divisors of a number? Actually this problem has already been met in testing prime numbers. Suppose we start with a number X then, obviously, 1 and X are divisors. Now try to see if 2 divides X – if so then 2 and $X/2$ are a pair of divisors. We can repeat the process with 3, 4 and so on – but how far do we need to go? We want to stop at the last pair N and X/N for which $N^2 \leqslant X$. If $N^2 = X$ then X is a perfect square and N is a repeated divisor which we will only count once. Thus a perfect square has an odd number of divisors. We are in a position, then, to produce a simple program (16.3) to produce some evidence. We assume that we already know the divisors of 1, 2 and 3.

Program 16.3 – collecting data about divisors

```
100   REM PROGRAM 16.3 – DIVISORS – DATA COLLECTION
110   FOR X = 4 TO 100
120      T = 2
130      N = 2
140      D = X/N
150      IF D = INT(D) THEN T = T + 2
```

```
160      IF N*N = X THEN T = T - 1
170      N = N + 1
180      IF N*N <= X THEN GOTO 140
190      PRINT X, T
200  NEXT X
```

This, then, produces a table giving the number of divisors for each integer between 4 and 100 and can easily be modified to produce more. To find the smallest number with *T* divisors we look down the right-hand column until we find the first occurrence of the number *T* and look at the corresponding entry of *X* in the left-hand column. With a little modification we can make the computer do this for us. In order to tabulate the smallest values of *X* corresponding to values of *T* from 3 to 20, say, we need to introduce an array which keeps track of whether a smallest number for a given *T* has yet been found. We also need to keep a count of how many of the elements of the array have been found so that we can determine when to stop scanning through values of *X*.

We can use the procedure of Program 16.3 with some changes at the beginning and end. Suitable amendments to that program are given in Program 16.4.

Program 16.4 – Further evidence about divisors

```
 50 REM PROGRAM 16.4 – DIVISORS – PRODUCE EVIDENCE
 60 DIM D(20)
 70 FOR I = 3 TO 20
 80    D(I) = 0
 90 NEXT I
100 F = 0
110 X = 4
120 T = 2
130 N = 2
140 D = X/N
150 IF D = INT(D) THEN T = T + 2
160 IF N*N = X THEN T = T - 1
170 N = N + 1
180 IF N*N <= X THEN GOTO 140
190 IF T > 20 THEN GOTO 240
200 IF D(T) > 0 THEN GOTO 240
210 D(T) = X
220 F = F + 1
230 IF F = 18 THEN GOTO 260
240 X = X + 1
250 GOTO 120
260 FOR I = 3 TO 20
270    PRINT I, D(I)
280 NEXT I
```

In Program 16.4 lines 120 to 180 are the procedure from Program 16.3 to find the number of divisors of X. If it has 20 or less divisors we look in the appropriate element of the array D(T) to see if a smaller value of X with T divisors has already been found. If it has not then we set D(T) to the current value of X and increment F which is counting the number of elements of the array that have been filled. When F reaches 18 the array has been filled and the process terminates.

Undoubtedly this program would be more elegant (and transparent) if structures such as REPEAT ... UNTIL and IF ... THEN ... ELSE had been used. If the reader has access to a microcomputer (such as the BBC micro) that allows such structures it might be a useful exercise to make this program more tidy.

This program may, however, take a very long time before any results appear on the screen. It might be advisable to reduce the range of T from 20 down to 12, say, just to prove that the program is working correctly. On most microcomputers you can interrupt a program using a suitable key (STOP, BREAK, ESCAPE etc.) and just use a *direct* command such as PRINT X,T,F or FOR I = 3 TO 20 : PRINT D(I) : NEXT I. Provided the program has not been altered nor new variables introduced then typing the phrase CONT should make the program resume from the point at which it was interrupted. Eventually we end up with table 16.1. Is there any discernible pattern in the X column? The explanation of why the program takes so long is now obvious. Is it just a coincidence

Table 16.1

T	X	T	X
3	4	12	60
4	6	13	4096
5	16	14	192
6	12	15	144
7	64	16	120
8	24	17	65536
9	36	18	180
10	48	19	262144
11	1024	20	240

that the large values of X appear opposite ***prime*** values of T? A reasonable hypothesis would be that if T is a prime number then the smallest number with T divisors is 2^{T-1}. If we accept this then we need only investigate non-prime values of T and we can add a subroutine to Program 16.4 to fill in $D(T)$ with 2^{T-1} for any prime values of T in the range being used. Thus suitable amendments to Program 16.4 are:

```
 105 GOSUB 1000
 290 END
1000 D(3) = 4
1010 FORT = 4 TO 20
```

```
1020      P = 2
1030      R = T/P
1040      IF R = INT(R) THEN GOTO 1120
1050      P = P + 1
1060      IF P*P <= T THEN GOTO 1030
1070      D(T) = 1
1080      FOR I = 1 TO T - 1
1090        D(T) = D(T)*2
1100      NEXT I
1110      F = F + 1
1120  NEXT T
1130  RETURN
```

The loop from 1070 to 1100 is to calculate 2^{T-1} exactly since 2↑(T - 1) will not always produce an integer. In this case, though, it is not particularly important. If your micro allows *integer* variables (usually with names such as X%, T% etc.) the program may well work faster if they are used.

With the evidence now available we can see that 2 and 3 are frequently occurring factors of the values of X and so it would be a reasonable step to explore the factors of X further (see Table 16.2). Is it obvious that a number

Table 16.2

T	X	Factors
2	2	2^1
3	4	2^2
4	6	2^1 3^1
5	16	2^4
6	12	2^2 3^1
7	64	2^6
8	24	2^3 3^1
9	36	2^2 3^2
10	48	2^4 3^1
11	1024	2^{10}
12	60	2^2 3^1 5^1
13	4096	2^{12}
14	192	2^6 3^1
15	144	2^4 3^2
16	120	2^3 3^1 5^1
17	65536	2^{16}
18	180	2^2 3^2 5^1
19	262144	2^{18}
20	240	2^4 3^1 5^1

such as $2^2.3^1.5^1$ has 12 divisors? The possible divisors of 2^2 are $(1,2,2^2)$, those of 3^1 are $(1,3)$ and those of 5^1 are $(1,5)$. Thus any divisor of 60 can be found by multiplying together one of the numbers from each of the three brackets. This can be done in $3 \times 2 \times 2$ different ways confirming that 60 has, indeed, 12 divisors. We can generalise this result. Suppose $p_1, p_2, p_3, \ldots$ are different prime numbers and that $a, b, c, \ldots$ are natural numbers. Then p_1^a has $(a+1)$ divisors, $p_1^a.p_2^b$ has $(a+1)(b+1)$ divisors $p_1^a.p_2^b.p_3^c$ has $(a+1)(b+1)(c+1)$ divisors etc. Also, within the set of numbers of the form p_1^a, 2^a is obviously the smallest. Suppose $a \geqslant b \geqslant c \geqslant \ldots$, then the smallest number of the form $p_1^a.p_2^b$ is $2^a.3^b$ and similarly the set $\{p_1^a.p_2^b.p_3^c : p_1, p_2, p_3 \text{ prime}, a, b, c \text{ integer}, a \geqslant b \geqslant c > 0\}$ has $2^a.3^b.5^c$ as its smallest element.

Now consider $T = 15$ – this can be factorised as 15×1 or as 5×3. Thus numbers with 15 divisors must be either of the form p_1^{14} or $p_1^4.p_2^2$. There are, then, just two candidates to be considered for the smallest number with 15 divisors, viz. 2^{14} or $2^4.3^2$ and there is no need to evaluate either of these to make the comparison since 2^{10} is clearly much greater than 3^2. Hence $X = 2^4.3^2 = 144$ is the smallest number with 15 divisors, which agrees with the result in the tables. Similarly for $T = 16$ the possible factorisations written in ascending order are 16, 8×2, 4×4, $4 \times 2 \times 2$, $2 \times 2 \times 2 \times 2$ and these give the following candidates for X: 2^{15}, $2^7.3$, $2^3.3^3$, $2^3.3.5$ and $2.3.5.7$. Clearly $2^{15} > 2^7.3$ (since $2^8 > 3$), similarly $2^7.3 > 2^3.3^3$ (since $2^4 > 3^2$) and $2^3.3^3 > 2^3.3.5$ (since $3^2 > 5$) but $2^3.3.5 < 2.3.5.7$ (since $2^2 < 7$) and so $X = 2^2.3.5 = 120$ is the required solution.

Can you generalise this process into an algorithm and so produce a program which will find the smallest number X with a given number T of divisors?

16.3 THE MAGIC SQUARE REVISITED

In Chapter 2 a problem concerning magic squares was introduced. Given a 6 by 6 magic square containing the numbers 1 to 36, each used just once, the sum of each row and column must be $6 \times (1 + 36)/2 = 111$. In how many different ways can 6 different numbers be taken from the 36 numbers 1 to 36 such that they total 111?

The simplest kind of program (16.5) to tackle this problem uses nested loops to hunt for sextuplets of the form (I, J, K, L, M, N) in ascending order which total 111. Depending upon the type of microcomputer this program takes typically around 15 hours to produce a solution! If we look at the order in which the sextuplets are taken it is not quite so surprising. They start:

$$(1,2,3,4,5,6)\ (1,2,3,4,5,7) \ldots (1,2,3,4,5,36)$$
$$(1,2,3,4,6,7)\ (1,2,3,4,6,8) \ldots (1,2,3,4,6,36)$$
$$(1,2,3,4,7,8)\ (1,2,3,4,7,9) \ldots (1,2,3,4,7,36)$$

and it is not until (1,2,3,34,35,36) that this first success is scored. Similarly, although (16, 17, 18, 19, 20, 21) is the last success, the program still hacks away to the bitter end at (31,32,33,34,35,36). In fact the algorithm searches through all the combinations of 6 items from 36 and so, from the appropriate binomial coefficient, we see that the test in line 180 is performed

$$\frac{36 \times 35 \times 34 \times 33 \times 32 \times 31}{1 \times 2 \times 3 \times 4 \times 5 \times 6} = 1\,947\,792 \quad \text{times!}$$

That test also involves some 10 million additions so the program really is not very clever.

Program 16.5 – A first try at the magic square problem

```
100  REM PROGRAM 16.5 – MAGIC SQUARE VERSION 1
110  S = 0
120  FOR I = 1 TO 31
130     FOR J = I + 1 TO 32
140        FOR K = J + 1 TO 33
150           FOR L = K + 1 TO 34
160              FOR M = L + 1 TO 35
170                 FOR N = M + 1 TO 36
180                    IF I + J + K + L + M + N = 111 THEN S = S + 1
190                 NEXT N
200              NEXT M
210           NEXT L
220        NEXT K
230     NEXT J
240  NEXT I
250  PRINT S
```

Suppose we just consider the upper limits for each of the loop variables I, J, K, L, M, N. Obviously there is no point in letting I, the smallest of these, pass 16 – since $17 + 18 + 19 + 20 + 21 + 22 > 111$. Is it possible to find an upper limit for J? J will be biggest when I is smallest so consider the sextuplet: (1, J, J + 1, J + 2, J + 3, J + 4) which totals $5J + 11$. If this is not to exceed 111 we have the inequality $5J + 11 \leqslant 111$ for which $J \leqslant 20$. Similarly an upper limit for K is found from (1, 2, K, K+1, K+2, K+3) which yields $4K + 9 \leqslant 111$ and so $K \leqslant 25$. We know that L can be as large as 34, M as large as 35 and N as large as 36. However, we can remove the innermost loop altogether. Once the 5 quantities I, J, K, L and M are known then we can find N by subtracting their total from 111. If N is both greater than M and less than 37 then we have a successful sextuplet.

Program 16.6 produces a successful result in around 2 hours. By setting a running total T we can check how many combinations are hunted through this time. Inserting 115 T = 0, 185 T = T + 1 and 230 PRINT T we find that the test at line 180 is performed 269 084 times confirming that Program 16.6 is about 8 times faster than Program 16.5. We can still go further in improving the efficiency of the algorithm. The upper bound of 20 on J is only a *least* upper bound on J when $I = 1$. In general we want the sextuplet: (I, J, J + 1, J + 2, J + 3, J + 4) to have a total no greater than 111 and so we have the inequality $I + 5J + 10 \leqslant 111$ for which $J \leqslant (101 - I)/5$ and this gives the least upper bound for J as a function of I. Using a similar argument on (I, J, K, K + 1, K + 2, K + 3) we have that $K \leqslant$

Program 16.6 – A better attempt at the magic square problem

```
100 REM PROGRAM 16.6 – MAGIC SQUARE – VERSION 2
110 S = 0
120 FOR I = 1 TO 16
130    FOR J = I + 1 TO 20
140       FOR K = J + 1 TO 25
150          FOR L = K + 1 TO 34
160             FOR M = L + 1 TO 35
170                N = 111 - (I + J + K + L + M)
180                IF (N > M) AND (N < 37) THEN S = S + 1
190             NEXT M
200          NEXT L
210       NEXT K
220    NEXT J
230 NEXT I
240 PRINT S
```

$(105 - I - J)/4$ which gives the least upper bound on K as a function of both I and J. A similar bound for L can be found as a function of I, J and K. Turning now to the lower bounds can we find *greatest* lower bounds for the loop variables? Obviously there are many solutions for which I, J, and K are consecutive such as (1, 2, 3, 34, 35, 36) and (2, 3, 4, 33, 34, 35). However, not until $I = 9$ can I, J, K and L be consecutive. For general I, J, K the value of L will be smallest when the remaining two elements are largest, hence we consider the sextuplet (I, J, K, L, 35, 36). We seek the smallest value of L for which the total is 111 or more. Thus $I + J + K + L + 71 \geqslant 111$ which gives $L \geqslant 40 - I - J - K$ as a greatest lower bound on L as a function of I, J and K. However, this value of $40 - I - J - K$ may yield a value for L which is smaller than $K + 1$ and which would break the rule about I, J, K, L, M, N being in ascending order. Hence the greatest lower bound of L is the greater of $40 - I - J - K$ and $K + 1$. Using a similar argument we find that the greatest lower bound on M is the greater of $75 - I - J - K - L$ and $L + 1$. Now using these tighter restrictions on I, J, K, L and M we have cast out all the fruitless initial 5 values of the sextuplets. Hence each legal value of I, J, K, L, M will have an associated unique N to produce the required total of 111 and so we can even dispense with the M loop entirely. Suppose $MU = INT((110 - I - J - K - L)/2)$ is the least upper bound on M and $ML = \max((75 - I - J - K - L), (L + 1))$ is its greatest upper bound then the M loop would have produced $MU - ML + 1$ successes. Using all these refinements we have a greatly improved algorithm (Program 16.7).

This program now takes around 10 minutes to produce an answer. With a little extra knowledge about how the BASIC language operates it is possible to code this algorithm in a more efficient way. In particular if integer arithmetic and variables are available then considerable savings in time can be made. In general, programs will run faster if constants such as 101, 105, 108, 110, 1, 2, 3, 4, 5 that appear in the program are stored, initially, in variables as each time BASIC comes across a constant it has to check that it is grammatically correct (for example, does not contain 2 decimal points or stray alphabetic characters). Again some of the arithmetic, such as 40 – I – J – K and 75 – I – J – K – L can be spread outside the loops. Using these refinements on a BBC microcomputer our fastest time so far is 54 seconds! Thus with a little mathematical argument concerning integers we have been able to reduce the running time of the program from about 16 hours to under one minute, that is, a reduction by a factor of about 1000.

Program 16.7 – A much more efficient attack on the magic square

```
100  REM PROGRAM 16.7 – MAGIC SQUARE – VERSION 3
110  S = 0
120  FOR I = 1 TO 16
130     FOR J = 1 + 1 TO (101 - I)/5
140        FOR K = J + 1 TO (105 - I - J)/4
150           LL = 40 - I - J - K : IF K + 1 > LL THEN LL = K + 1
160           FOR L = LL TO (108 - I - J - K)/3
170              ML = 75 - I - J - K - L : IF L + 1 > ML THEN
                 ML = L + 1
180              MU = INT ((110 - I - J - K - L)/2)
190              S = S + MU - ML + 1
200           NEXT L
210        NEXT K
220     NEXT J
230  NEXT I
240  PRINT S
```

16.4 A SQUARE SHUFFLE

The following problem is concerned with defining a shuffle on a square array of coloured bricks and then determining for a given size of array the minimum number of shuffles required to bring the array back to its initial state. The investigation is described in the Leapfrogs booklet *Moves* and there is a film *Square Shuffle* to accompany it. The problem was used as the basis of a workshop during an Association of Teachers of Mathematics conference and the following section is taken from an article 'Scampi and Chips – Computers on the Mathematics Menu' by Afzal Ahmed and Adrian Oldknow which appeared in *Mathematics Teaching* (No. 94, March 1981) and which is reprinted with kind permission of the Editor.

36 coloured blocks are placed in a 6 × 6 square; there are 9 each of Red, Green, Yellow, Blue and initially the Red blocks are in the top left-hand corner forming a 3 × 3 square and there is a similar Green square on the top right, Yellow square in the bottom left and Blue square in the bottom right. The shuffle consists of taking blocks away from the initial square four at a time from its outside and placing them in an adjacent 6 × 6 square, which is initially empty, as shown in Fig. 16.1.

```
R R R G G G
R R R G G G
R R R G G G
Y Y Y B B B
Y Y Y B B B
Y Y Y B B B

  R R G G
R R R G G G
R R R G G G
Y Y Y B B B
Y Y Y B B B                 R G
  Y Y B B                   Y B

  R R G G
  R R G G
R R R G G G                 R G
Y Y Y B B B                 Y B
  Y Y B B                   R G
  Y Y B B                   Y B

  R R G G                   R G
  R R G G                   Y B
  R R G G                   R G
  Y Y B B                   Y B
  Y Y B B                   R G
  Y Y B B                   Y B

    R G                     R G
  R R G G                   Y B
  R R G G                   R G
  Y Y B B                   Y B
  Y Y B B               R G R G
    Y B                 Y B Y B

    R G                     R G
    R G                     Y B
  R R G G               R G R G
  Y Y B B               Y B Y B
    Y B                 R G R G
    Y B                 Y B Y B
```

Fig. 16.1 (continued next page)

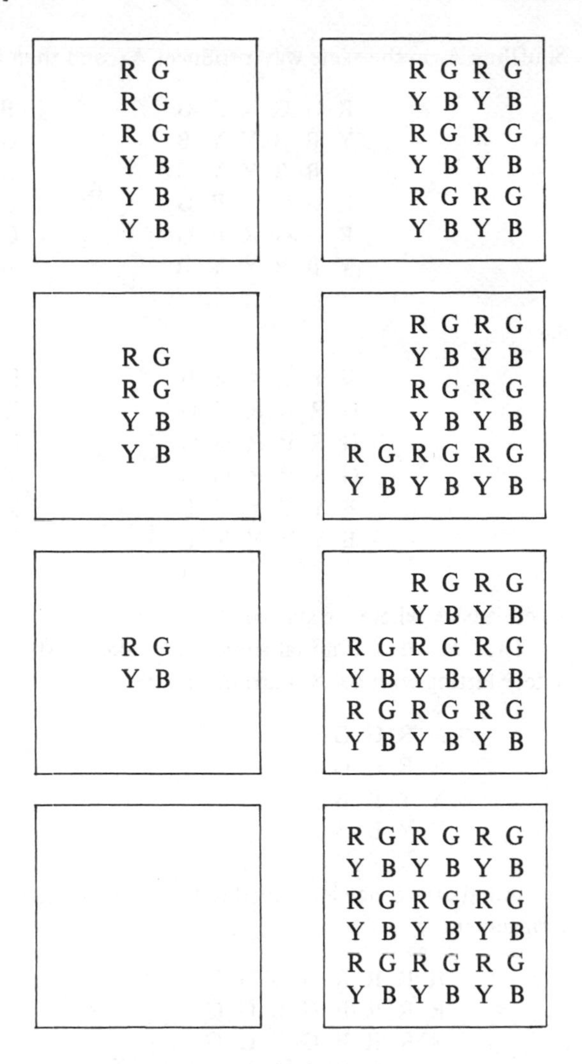

Fig. 16.1 (continued)

Thus after one shuffle the intitial array, A_0, becomes A_1.

$$A_0 = \begin{matrix} R & R & R & G & G & G \\ R & R & R & G & G & G \\ R & R & R & G & G & G \\ Y & Y & Y & B & B & B \\ Y & Y & Y & B & B & B \\ Y & Y & Y & B & B & B \end{matrix} \qquad A_1 = \begin{matrix} R & G & R & G & R & G \\ Y & B & Y & B & Y & B \\ R & G & R & G & R & G \\ Y & B & Y & B & Y & B \\ R & G & R & G & R & G \\ Y & B & Y & B & Y & B \end{matrix}$$

Shuffling A_1 in the same way produces A_2, and then in turn A_3,

$$A_2 = \begin{array}{cccccc} R & G & G & R & R & G \\ Y & B & B & Y & Y & B \\ Y & B & B & Y & Y & B \\ R & G & G & R & R & G \\ R & G & G & R & R & G \\ Y & B & B & Y & Y & B \end{array} \qquad A_3 = \begin{array}{cccccc} B & Y & B & Y & Y & B \\ G & R & G & R & R & G \\ B & Y & B & Y & Y & B \\ G & R & G & R & R & G \\ G & R & G & R & R & G \\ B & Y & B & Y & Y & B \end{array}$$

then ...

$$A_4 = \begin{array}{cccccc} B & Y & Y & Y & B & B \\ G & R & R & R & G & G \\ G & R & R & R & G & G \\ G & R & R & R & G & G \\ B & Y & Y & Y & B & B \\ B & Y & Y & Y & Y & B \end{array} \qquad A_5 = \begin{array}{cccccc} R & R & R & G & G & G \\ R & R & R & G & G & G \\ R & R & R & G & G & G \\ Y & Y & Y & B & B & B \\ Y & Y & Y & B & B & B \\ Y & Y & Y & B & B & B \end{array}$$

which is back where we started.

Thus our 6×6 initial square has taken 5 shuffles to return to its original state. Starting with a 4×4 arrangement

$$\begin{array}{cccc} R & R & G & G \\ R & R & G & G \\ Y & Y & B & B \\ Y & Y & B & B \end{array}$$

we find that we need 3 shuffles to return to the original, and with an 8×8 arrangement

$$\begin{array}{cccccccc} R & R & R & R & G & G & G & G \\ R & R & R & R & G & G & G & G \\ R & R & R & R & G & G & G & G \\ R & R & R & R & G & G & G & G \\ Y & Y & Y & Y & B & B & B & B \\ Y & Y & Y & Y & B & B & B & B \\ Y & Y & Y & Y & B & B & B & B \\ Y & Y & Y & Y & B & B & B & B \end{array}$$

we need just 4 shuffles to be back where we started.

The problem, then, is to find an expression for the minimum number of shuffles required to bring an $N \times N$ square (with N even) back to its original arrangement.

Obviously the physical job of shuffling blocks, tiles or counters to collect data is very time-consuming (and has only a vague connection with mathematical activity) and the actual amount of time spent 'mathematising' is correspondingly low, so this is an ideal situation for the computer to do the monotonous donkey-work of data collection, leaving us to do the hypothesis formulation and testing.

But how do we put the problem into a form that a computer can handle? The very decision to involve a computer at all forces a narrowing of the possible approaches to the problem and channels us in a particular direction imposed by the way that the computer stores its data.

The square will have to be held in the computer as an 'array' A, say, whose elements are labelled

A(1,1), A(1,2), A(1,3), A(1,4), A(1,5), A(1,6)
A(2,1), A(2,2), A(2,3), A(2,4), A(2,5), A(2,6)
A(3,1), ...
A(4,1), ...
A(5,1), ...
A(6,1), ...

Suppose, then, that the initial array A is shuffled into another array B, initially 'empty'. Then we have the mapping

A(1,1) → B(5,5)
A(6,1) → B(6,5)
A(1,6) → B(5,6)
A(6,6) → B(6,6)

corresponding to the first stage of the shuffle. If we do this for all the elements then we show the relationship between the A and B arrays as in Table 16.3.

Table 16.3

3,3	3,4	3,2	3,5	3,1	3,6
4,3	4,4	4,2	4,5	4,1	4,6
2,3	2,4	2,2	2,5	2,1	2,6
5,3	5,4	5,2	5,5	5,1	5,6
1,3	1,4	1,2	1,5	1,1	1,6
6,3	6,4	6,2	6,5	6,1	6,6

Thus looking at row 4, column 3 we see that B(4,3) is the element that was in A(5,2), and so on. The symmetry of this table is striking; reading the column numbers 'backwards' (from right to left) we have 6, 1, 5, 2, 4, 3 – and the same for row numbers.

Just double checking for the 8 × 8 array we confirm that the rows and columns read 'backwards' give 8, 1, 7, 2, 6, 3, 5, 4. So we are in good shape to seek the general rule: where does A(I, J) go?

Since both the row, I, and the column, J, indexes behave the same way the problem is considerably reduced (Fig. 16.2).

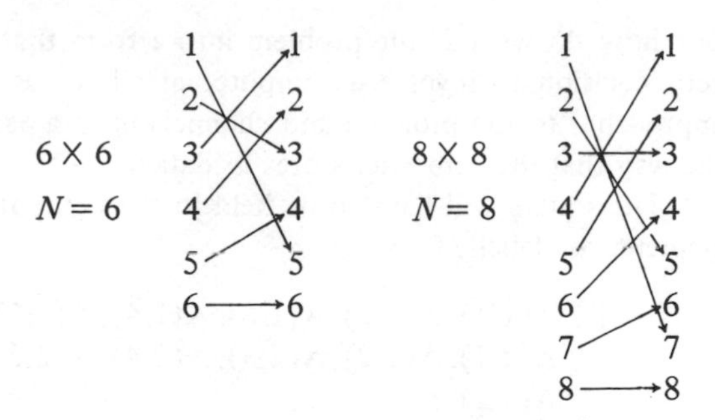

Fig. 16.2

The mapping, then, is best considered in two halves. If I is in the top half of the domain then the mapping $1 \rightarrow N-1$, $2 \rightarrow N-3$, etc. generalises to $I \rightarrow N+1-2I$; and if I is in the bottom half of the domain then the mapping $N/2+1 \rightarrow 2$, $N/2+2 \rightarrow 4$, etc. generalises to $I \rightarrow 2I-N$, so we have the conditional mapping, for N and I integers, and with N even:

$$I \rightarrow \begin{cases} N+1-2I & \text{if} \quad 1 \leqslant I \leqslant N/2 \\ 2I-N & \text{if} \quad N/2 < I \leqslant N \end{cases}$$

Just to check, consider A(2,5) in the 6×6 case. $N = 6$, and so $N/2 = 3$, $2 < 3$ and thus $2 \rightarrow N+1-2(2) = 3$ but $5 > 3$ and so $5 \rightarrow 2(5)-N = 4$ and we can expect A(2,5) to map B(3,4) which we can check by looking at row 3, column 4 of the earlier table.

Now we have defined the general shuffle an initial algorithm is straightforward.

1. Define an $N \times N$ array A (with N even) and fill its top left-hand quarter with 1s, its top right-hand quarter with 2s, its bottom left-hand with 3s, and its bottom right-hand corner with 4s. Set a count, K, to zero. Define an $N \times N$ array B.
2. Fill B with the corresponding elements of A using the generalised shuffle rule and add 1 to the count.
3. Check whether: all elements in the top left quarter of B are 1s *and* all elements of top right quarter are 2s *and* all the bottom left are 3s *and* all the bottom right are 4s. If test is successful then EXIT with K as the required value. Otherwise copy B into A (set $A(I,J) = B(I,J)$ for $I = 1, \ldots, N$ and $J = 1, \ldots, N$) and go to step 2.

This finds the number of shuffles, K, needed to return any $N \times N$ array (N even) back to its starting configuration. It is also simple to program – and, in particular, if we put the whole procedure inside a loop that makes N take the values 2, 4, 6, 8, ... then we have the desired data-producer (Program 16.8).

Program 16.8 – A first try at the square shuffle

```
100 REM PROGRAM 16.8 – SQUARE SHUFFLE 1
110 DIM A(20,20), B(20,20)
120 FOR M = 1 TO 10
```

```
130       N = 2*M
140       REM FILL THE ARRAY
150       FOR I = 1 TO M
160         FOR J = 1 TO M
170           A(I,J) = 1 : A( I,J +M) = 2
180           A(I +M,J) = 3 : A(I +M,J +M) = 4
190         NEXT J
200       NEXT I
210       K = 0
220       REM NOW SHUFFLE A INTO B
230       FOR I = 1 TO M
240         FOR J = 1 TO M
250           I1 = 2*I : J1 = 2*J
260           I2 = N + 1 - I1 : J2 = N + 1 - J1
270           B(I2,J2) = A(I, J) : B(I2,J1) = A(I,J +M)
280           B(I1,J2) = A(I + M,J) : B(I1,J1) = A(I + M, J +M)
290         NEXT J
300       NEXT I
310       K = K +1 : F = 0
320       REM NOW TEST IF B IS DIFFERENT FROM A
330       FOR I = 1 TO M
340         FOR J = 1 TO M
350           IF B(I,J) = 1 AND B(I,J +M) = 2 AND B(I +M,J) = 3
              AND B(I +M,J +M) = 4 THEN GOTO 370
360         F = 1 : I = M : J = M
370         NEXT J
380       NEXT I
390       IF F = 0 THEN GOTO 470
400       REM REFILL A FROM B
410       FOR I = 1 TO N
420         FOR J = 1 TO N
430           A(I, J) = B(I, J)
440         NEXT J
450       NEXT I
460       GOTO 220
470       REM SUCCESS
480       PRINT N, K
490   NEXT M
500   END
```

Line 360 of Program 16.8 is included to illustrate a technique to avoid 'jumping out of a loop', which is necessary to use with some versions of BASIC.

Unfortunately (or otherwise) that is far from the end of the story. The program ran nicely to start with, confirming the values we already knew for $N = 2, 4, 6, 8$ – but then it began to get more and more ponderous between outputs. With a little thought it is not surprising. Consider the 10×10 case, for which K turns out to be 9. For each shuffle we have to fill $N^2 = 100$ elements of B then test any number between 1 and 100 elements of B looking for a mismatch between it and the original configuration and, if unsuccessful, then fill 100 elements of A. So we need 200 assignments and quite a few tests at each pass of the algorithm – and there are 9 passes in all. So, in order to make the procedure more efficient some problem-reduction is called for. The next, small, step taken was to argue that we need only keep track of the elements originally in the top left-hand corner (the Reds, or 1s), and claim that when they found their way back to their starting point again all the other elements would also have done so. This intuitive reduction caused some fierce argument as no simple proof of the proposition could be found and merely showing that the values of K output by the algorithm duplicated the known values was hardly convincing to the sceptical. The only change in the original algorithm was to put 0s instead of 2s, 3s and 4s in the starting values of A and to test when the upper left-hand corner of B contained all 1s. As the improvement in performance of the algorithm was not sufficient to stop huge pauses at times, for example while it found $K = 23$ for $N = 24$, the argument was really in vain.

Apart from the slowness of the procedure for large N there is also, of course, an increasing demand for storage in the computer – the 2 $N \times N$ matrices, A and B, require around $10N^2$ bytes of storage using, say, BASIC on a micro that takes 5 bytes to store each variable, and so, for example, an 8K PET (or 16K 380Z with BASIC loaded) cannot cope with $N=30$. So a very much more far-reaching reduction of the problem is called for. Therefore another, analogous, problem was considered. Look at the indexes that, say, A(2,5) (in the case $N = 6$) runs through before returning to its starting point:

$$(2,5) \rightarrow (3,4) \rightarrow (1,2) \rightarrow (5,3) \rightarrow (4,1) \rightarrow (2,5) \ .$$

We could, then, consider each element $A(I,J)$ and find how many steps it took to get back to its starting position. If we do this for the 6×6 case we find that *every* element takes 5 steps except A(6,6) which takes just one. In the 8×8 case most elements take 4 steps, some take 2 (such as A(2,5)) and there are 4 elements (A(8,8), A(3,3), A(3,8) and A(8,3)) which take just 1 step. So perhaps it is a good time to introduce the idea of the *period* of an element, that is the smallest number of steps it takes to get back to its starting position, and so make the claim that K, the smallest number of shuffles required to bring the array A back to its starting position, is the least common multiple of the periods of its elements. But before rushing back to the computer it is worthwhile considering the period of an individual element more closely. Consider A(6,5) in the 8×8 case.

$$(6,5) \rightarrow (4,2) \rightarrow (1,5) \rightarrow (7,2) \rightarrow (6,5)$$

contains the 'I-chain'

$$\underset{\uparrow\_\_\_\_\_\_\_\_\_\_\_\_\_\_\_\_\_\_\_\_\_\_\_\_\_\_\_\_\_\_\_\_\_\_\_\_|}{6 \rightarrow 4 \rightarrow 1 \rightarrow 7}$$

and the 'J-chain'

$$\underset{\uparrow\_\_\_\_|}{5 \rightarrow 2}$$

The period of the *I*-chain is 4 and that of the *J*-chain is 2 and so the period of A(6, 5) is 4 which is the least common multiple of 4 and 2.

So perhaps we ought to define another term, the ***trajectory*** of an element. Consider a mapping such as shown in Fig. 16.3 and for each element consider the chain of its images,

$$1 \to 2 \to 3 \to 1 \ldots$$
$$2 \to 3 \to 1 \to 2 \ldots$$
$$3 \to 1 \to 2 \to 3 \ldots$$
$$4 \to 5 \to 5 \ldots$$
$$5 \to 5$$
$$6 \to 6$$

which we can illustrate as

$$1 \to 2 \to 3 \quad 4 \to 5 \quad 6\ .$$

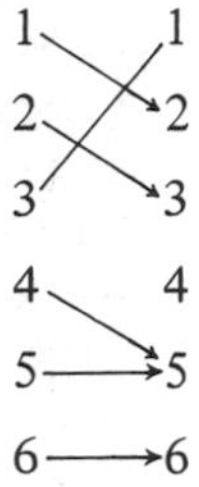

(with return arrows from 3 back to 1, from 5 to itself, and from 6 to itself)

1 → 2, 2 → 3, 3 → 1, 4 → 5, 5 → 5, 6 → 6

Fig. 16.3

The trajectory of the element under the mapping f is the set $\{x, f(x), f(f(x)), \ldots\}$ and thus the trajectories of the elements 1, 2 and 3 in our example are all identical, {1, 2, 3}; the trajectory of 4 is {4, 5}; that of 5 is {5} and that of 6 is {6}. An element that is the only member of its own trajectory, like 5 and 6, is an ***invariant element*** and if the trajectory of an element is like that of 1, 2 and 3 then it is a ***cycle.*** For the mappings that we have looked at arising from the square shuffle we see that any element is either an invariant element, or its trajectory is a cycle. In the 8 × 8 case we have the cycle

$$1 \to 7 \to 6 \to 4$$

(with a return arrow from 4 back to 1)

of length 4, the cycle

$$2 \to 5$$

(with a return arrow from 5 back to 2)

of length 2 and the two invarient elements 3 and 8. Thus we expect that the period of an element A(*I*, *J*) will be the least common multiple of the lengths of its *I*-cycle and its *J*-cycle. Armed with this and the earlier claim about the relation of *K*, the length of a shuffle, to the periods of the elements A(*I*, *J*), we have all the information we need just by considering the trajectories of elements under the 'fundamental' mapping,

$$I \rightarrow \begin{cases} N+1-2I & \text{if} \quad 1 \leqslant I \leqslant N/2 \quad I, N \text{ integer} \\ 2I-N & \text{if} \quad N/2 < I \leqslant N \quad N \text{ even} . \end{cases}$$

So our third algorithm is to output the lengths of the distinct trajectories of the fundamental mapping for a given value of N.

1. Define a list A of N items and fill it with zeroes.
2. Scan the list to find the first zero element $A(I)$. If there are none then EXIT.
3. Set $A(I)$ to 1, set J to equal I, and a count, C, to zero.
4. Find the image J' of J under the fundamental mapping. If $J' = I$ then print C and go to step 2. Otherwise set $A(J')$ to 1, add 1 to the count, set J to equal J' and go to step 4.

Program 16.9 – The square shuffle tamed

```
100 REM PROGRAM 16.9 – SQUARE SHUFFLE 2
110 DIM A(200)
120 FOR N = 2 TO 200 STEP 2
130    FOR I = 1 TO N
140       A(I) = 0
150    NEXT I
160    PRINT N,
170    FOR I = 1 TO N
180       IF A(I) <> 0 THEN GOTO 290
190       A(I) = 1 : C = 0 : J = I
200       J2 = J*2
210       IF J2 > N THEN GOTO 240
220       J = N + 1 - J2
230       GOTO 250
240       J = J2 - N
250       C = C + 1
260       A(J) = 1
270       IF J <> I THEN GOTO 200
280       PRINT C;
290    NEXT I
300    PRINT
310 NEXT N
```

The first few results are shown in Table 16.4. This produces the conjecture that the length of each trajectory is a factor of the longest trajectory, for each N. (We were not doing group theory, were we?) It also looks as if $I = 1$ is always a member of the longest trajectory.

Given our earlier claims about the relationship of the lengths of trajectories to numbers of shuffles we are just considering least common multiples of common factors, so we have the conjecture that the number of shuffles required to put the square back to its original position is just the length of the longest trajectory under the corresponding fundamental mapping. What we have not managed to do is to find the relationship between N and K – but only a pretty efficient procedure to find K given N.

In all this, through, we have changed the original problem; harking back to the 6 X 6 case the reds, greens, yellows and blues have been forgotten and instead we have found how long it takes for each element to get back to its starting position. It just so happens that the answer turns out to be the same, but could there be a case where all the reds, greens, yellows and blues had returned to their respective corners but in which, say, some of the reds were permuted among themselves? In this case the value of K would be smaller than we propose (and, very probably, a factor of it), and yet intuition and evidence – but no proof – says that no such case exists!

The role of the computer, here, has been to force us to seek generalised rules at an earlier stage than would have been the case with 'hand' data collection. In doing so we have been forced to *use* algebra to express those rules in a form the computer can handle. By coming across problems of computer storage and calculation time we have had to consider not just the formulation of an algorithm, but also its efficiency, and this has led to further mathematical activity in *analysing* the efficiency of an algorithm as well as in seeking improved algorithms. What is has not done is to make *proof* any easier or less important, but merely to provide more evidence for conjectures than would otherwise not have been feasible to collect.

Table 16.4

N												
2	1	1										
4	3	1										
6	5	1										
8	4	2	1	1								
10	9	1										
12	11	1										
14	9	3	1	1								
16	5	5	5	1								
18	12	3	2	1								
20	12	6	1	1								
22	7	7	7	1								
24	23	1										
26	8	4	8	1	1							
28	20	5	2	1								
30	29	1										
32	6	6	6	3	3	6	1	1				
34	33	1										
.....												
64	7	7	7	7	7	7	7	7	7	1		
.....												
86	9	9	9	9	9	9	9	9	3	9	1	1

16.5 ACKERMANN STEERING

Anyone familiar with Meccano or Lego will probably have tried to make a vehicle that steers. The usual mechanism for a model is based upon a parallelogram but it soon becomes obvious that it has certain defects. Just as a differential is needed because the speeds of the driven wheels when rounding a bend are different so a mechanism is required to make the steered wheels point at different angles. Consider a car rounding a bend. Ideally the paths of all four wheels will be arcs of concentric circles. Suppose the front inner wheel FI is turned at an angle AI to the length of the car and the front outer wheel FO is turned at an angle AO. Then the perpendicular to each wheel should meet the line RO to RI of the rear wheels produced at a single point C. Suppose the transverse distance W between the front wheels is also the distance between the rear wheels and that L is the lengthwise distance between the two pairs of wheels. If the distance from RI, the rear inner wheel, to C, the centre of rotation, is R then we have: $L = R \tan(AI)$ and $L = (R + W) \tan(AO)$ from which we can eliminate R to give

$$\frac{W}{L} = \cot(AO) - \cot(AI) \ldots \qquad (16.2)$$

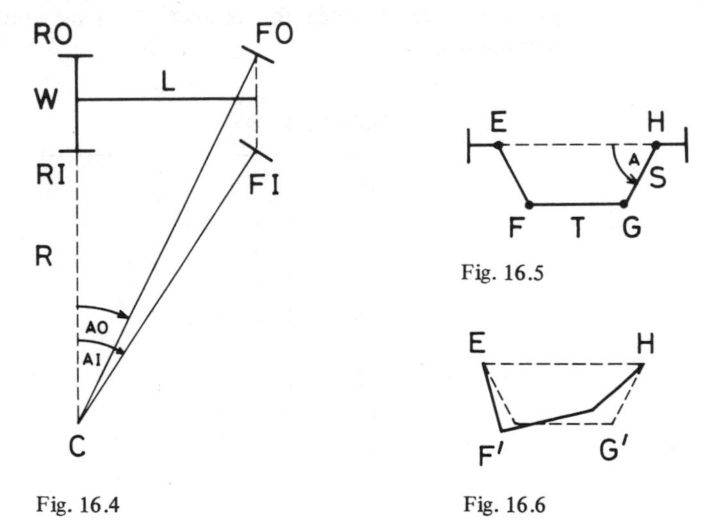

Fig. 16.4

Fig. 16.5

Fig. 16.6

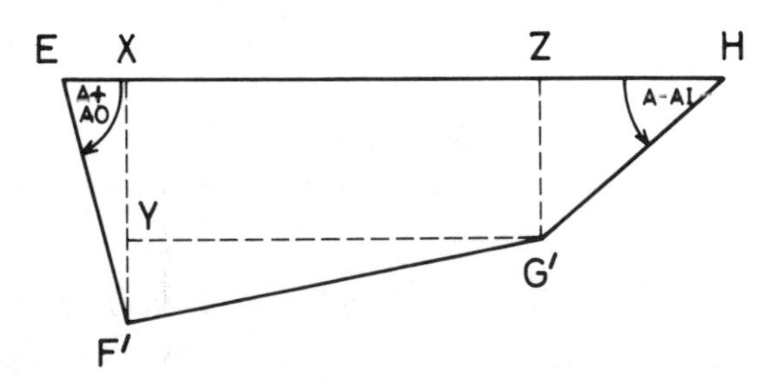

Fig. 16.7

This, then, enables us to find values of the angle AO for given values of AI. A trapezium is used as the basis for the Ackermann steering mechanism. The wheels are mounted on stub axles which pivot at points E and H and which are rigidly attached to rods EF and HG each of which have the same length S and, when the wheels are in line, make an angle A with the line of axles. The ends of these links are joined by a bar FG of length T freely pivoted at F and G. As FG $= W$ we have that the initial angle A of the trapezium satisfies

$$T + 2S\cos A = W \qquad (16.3)$$

If the wheel at H is the front inner wheel which is turned through an angle AI and the wheel at E is the front outer wheel which is turned through an angle AO we should be able to find a relation connecting AO, AI and the dimensions of the trapezium. From 16.7 we can divide the deformed trapezium into right-angled triangles to give:

$$(F'G')^2 = (F'Y)^2 + (YG')^2 = (F'X - G'Z)^2 + (EH - EX - ZH)^2$$

Thus

$$T^2 = (S\sin(A + AO) - S\sin(A - AI))^2 + + (W - S\cos(A + AO) - S\cos(A - AI))^2 \qquad (16.4)$$

Although this expression can be simplified somewhat it cannot be rearranged to give, say, AO as an explicit function of AI which could be compared with (16.2).

Ideally, then, for given values of W and L we would like to determine suitable defining values T, S and A for the appropriate trapezoidal steering gear that would ensure that the wheels took paths that were always arcs of concentric circles. However, if there were no such ideal choice then we would like to know the values of T, S and A to use that most closely approximate the desired relationship.

Since there is a relationship (16.3) connecting T, S and A there are, for given W and T, just two independent variables. Obviously from physical design constraints we do not want the steering gear to extend too far behind the front wheels so we are interested in cases in which $0 < T < W$ and, say, $0 < S < L/3$. For choices of T and S in these ranges we can compute A from $A = \cos^{-1}((W-T)/2S)$. For the chosen dimensions L, W of the car and T, S of the steering gear we can study the results produced over a range of values for AI, the angle of the inner wheel. For example we can first compute the desired value of AO from $AO = \cot^{-1}((W/L) + \cot(AI))$. With these values for the angles AI and AO we can calculate the discrepancy D between the left- and right-hand sides of equation (16.4), that is

$$D = S^2(\sin(A + AO) - \sin(A - AI))^2 + + (W - S\cos(A + AO) - S\cos(A - AI))^2 - T^2 .$$

Here we now have D as a function of AI (since AO is computed from AI) for given T and S. We seek values of T and S that make D as small as possible over the range of possible angles AI. For a modern car, such as a Vauxhall Cavalier, typical dimensions are $W = 1.4$ m, $L = 2.6$ m and a smallest turning circle of

10m (outer wheels to outer wheels). From this we have that the maximum value of AI is given by $\tan AI = 2.6/(5-1.4) = 2.6/3.6$. So AI has a maximum value of about 36° or 0.625 radians.

A simple program to tabulate D over a range of values of AI for given values of T and S can now be easily produced (Program 16.10).

Program 16.10 – To find suitable dimensions for steering geometry

```
100 REM PROGRAM 16.10 – STEERING PROBLEM
110 W = 1.4 : L = 2.6 : AM = 0.625
120 INPUT T, S
130 A = ATN(SQR(4*S*S - (W - T)*(W - T))/(W - T))
140 FOR AI = 0.05 TO AM STEP 0.05
150   AO = ATN(1/(W/L + 1/TAN(AI)))
160   F = S*(SIN(A+AO) - SIN(A - AI))
170   B = W - S*(COS(A+AO) + COS(A - AI))
180   D = F*F + B*B - T*T
190   PRINT AI, AO, D
200 NEXT AI
```

For different values of T we can try different values of S to explore the behaviour of the function D from one direction. By reversing the roles of S and T we can explore D from the other direction. Rather than tabulate AO and D against a range of values of AI we could just store the maximum value of $ABS(D)$ over the range of values of AI. We then seek a pair of values for T and S which minimise this chosen criterion for closeness of fit, that is, max $\{ABS(D)\}$ where $0 \leqslant AI \leqslant AM$. Here we have chosen a ***minimax*** criterion of fit. There are, of course, others such as least squares that can be just as easily applied. By hunting for a minimum of the greatest value for $ABS(D)$ we find a very good fit when $T=1.220, S=0.259$.

This problem exemplifies a kind of 'real life' application of mathematics which would prove far too complex to tackle with most students without the aid of a computer but which can be explored with quite elementary programming skills.

17

Transformation geometry

Transformations of the coordinate plane feature in many elementary mathematics courses but practical experience of applying the transformations is limited by the means of carrying them out. The computer, especially if it has a graphics facility, provides us with the means of performing transformations without the need for repetitive calculations. The choice of parameters can still be within the students control so that judgement is needed as to precisely what movement will produce the desired effect. With a suitable computer program a great deal of experience can be built up in a short time. For those whose mathematical background is right for it, the development of the program can be a valuable exercise, but we believe that for most students it is the use that will be of benefit.

17.1 THE COORDINATE SYSTEM AND DRAWING THE AXES

To handle the elements of transformation geometry in a computer each point is represented by its coordinates and these may be regarded as the components of a vector. The transformations are effected by multiplying the vectors by a suitable matrix. In Chapter 5 the matrices are stored in arrays, but here we will be working in two dimensions and the 2×2 matrices contain so few elements that arrays are unnecessary. We will show how translations, reflections, rotations and enlargements may be performed, and draw a shape and offer transformations that may be performed on it.

If no graphics facility is available on the computer, images cannot be satisfactorily reproduced on the screen, and the results would have to be output as a series of coordinates. We will assume a graphics facility that contains a statement to move the graphics 'pen' to any point on the screen and will use MOVE X,Y for this. The statement DRAW X,Y will be used to draw a line from the previous point used to the point (X,Y). Implementations of BASIC vary in the words they use for these and our example programs will need to be changed to the version used by the reader's machine. We will assume further that the clearing of text is independent of the clearing of graphics and will use CLS for text and CLG for graphics.

Each machine has a coordinate system in a wide range, usually much too wide for our purpose, and referred to an origin at one corner of the screen. It is convenient to have an origin in the middle of the screen at point (XS,YS) and these values will need to be added to the real coordinates when plotting. A scaling factor SC is used to multiply the coordinates for plotting to keep the range of actual coordinates within reason. If the pixels of the computer are not square then we also need a squash factor SQ to compensate. SQ is often about 0.8 or 0.9 and multiplies the Y coordinate only. A typical move instruction to the point (X,Y) thus takes the form

MOVE XS + SC*X, YS + SQ*SC*Y

For example the BBC microcomputer uses coordinates 0 to 1279 for X and 0 to 1023 for Y. To put the origin in the centre of the screen put XS = 639 and YS = 511. The choice of SC = 50 gives a reasonable range of possible coordinates (see Fig. 17.1).

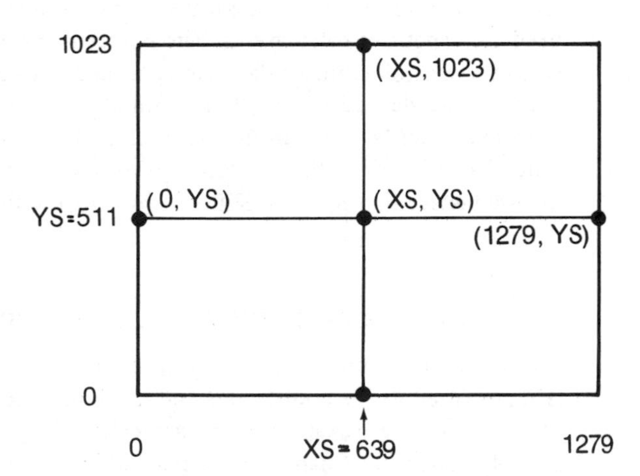

Fig. 17.1

We may start by drawing the axes with the program lines:

```
 93 MODE 4
100 REM**DRAW AXES**
105 CLS : CLG
110 XS = 639 : YS = 511 : SQ = 0.9 : SC = 50
115 MOVE XS,0 : DRAW XS, 1023
120 MOVE 0,YS : DRAW 1279,YS
```

It may also be desirable to add further program lines to put scales on the axes. Program 17.6 includes lines 125 to 155 for this (see Fig. 17.2). To move to the point (X, Y) we now use the statement

```
MOVE XS + SC*X, YS + SQ*SC*Y
```

and join this to the point (P, Q) we follow with

```
DRAW XS + SC*P, YS + SQ*SC*Q
```

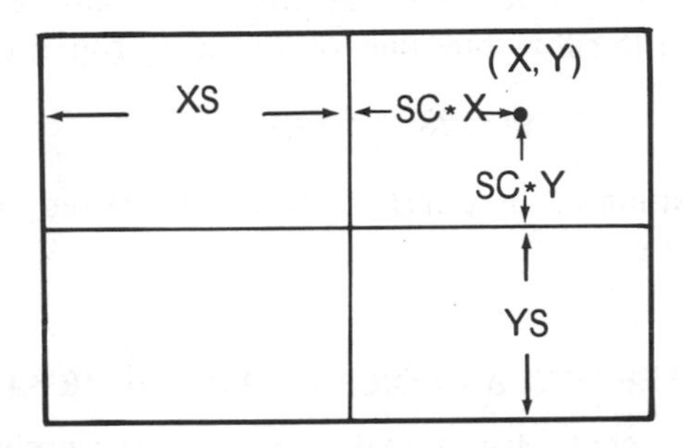

Fig. 17.2

17.2 INPUTTING A SHAPE

Having drawn the axes we are ready for the user to define a shape. The coordinates may be conveniently held in an array S in which row 1 contains the *x*-coordinates and row 2 the corresponding *y*-coordinates. Thus the *x*-coordinate of point 3 is in S(1,3) and the *y*-coordinate in S(2,3). In the example, the dimension line 205 allows for up to 10 points. The first point will be repeated at the end to simplify the drawing of a closed shape. The shape is drawn as the points are input and this helps the user to judge where the points are to go. To achieve this, line 255 moves the graphics 'pen' to the first point, but for subsequent points line 260 draws a line of the shape. The test in line 245 detects the input of (100, 100) that signals the end of the shape, and control branches to line 280 where the first point is copied into the end of the list. Line 285 then draws the final line of the shape.

```
200 REM**INPUT SHAPE**
205 DIM S(2,11)
215 CLS
220 PRINT "PLEASE TYPE COORDINATES OF SHAPE."
225 PRINT "AND 100,100 TO END."
230 N = 1
235 PRINT "POINT"; N;
240 INPUT S(1,N), S(2,N)
245 IF S(1,N) >= 100 THEN GOTO 280
250 IF N > 1 THEN GOTO 260
255 MOVE XS + SC*S(1, 1),YS + SQ*SC*S(2,1) : GOTO 265
260 DRAW XS + SC*S(1,N), YS + SQ*SC*S(2,N)
265 N = N + 1 : GOTO 235
280 S(1,N) = S(1,1) : S(2,N) = S(2,1)
285 DRAW XS + SC*S(1,N), YS + SQ*SC*S(2,N)
```

It should be noted that both text and graphics are required on the screen at the same time. This can be confusing unless steps are taken to keep them apart. One way is to define a text 'window' occupying a strip at the top or bottom of the screen. The instruction to do this may be inserted at line 95 and for the BBC computer to allow three lines of text at the top of the screen is

```
95 VDU28,0,2,39,0
```

The position of the x-axis will need to be moved to allow for this and YS given a different value.

17.3 OFFERING A CHOICE OF TRANSFORMATION

Having entered a shape into the computer we may give the student the opportunity of applying a transformation to the shape. To give flexibility to the program a choice of transformation may be offered in the form of a 'menu'. All that is required is to ask the user to choose a number and use it to transfer control to the appropriate program segment. Possible program lines for this are:

```
305 CLS
310 PRINT "TRANSLATE(1), REFLECT(2), ROTATE(3),"
311 PRINT "ENLARGE(4) OR END(5) -"
312 PRINT "CHOOSE A NUMBER";
315 INPUT CH
320 IF CH < 1 OR CH > 5 THEN GOTO 305
325 ON CH GOTO 400,500,600,700, 900
```

17.4 TRANSLATION

The simplest transformation is a translation and is specified by giving the x and y components of the translation vector. These components are then added to all the x and y coordinates of the shape to effect the translation. In Program 17.1

it is lines 400 to 435 that do this, the vector having components TX and TY. The loop in lines 425 to 435 effects the translation by adding TX to all the x-coordinates and TY to all the y-coordinates of the shape in array S. Next we draw the shape in its new position so that the student can see the effect of the translation he has chosen. We shall need to do this after other transformations too so the program section starting at line 800 does the drawing and may be used after any of the movements have been chosen. Line 810 move the graphics pen to the first point, then the loop in lines 815 to 825 joins each point to the next with the DRAW statement. Finally line 830 returns control to the menu so that another transformation may be selected.

17.5 REFLECTION

To reflect a point $P(x,y)$ in a line through the origin making an angle θ with the x-axis the coordinates of the image $Q(x',y')$ need to be calculated. To effect a reflection in the axis OA in the diagram we may make three movements (see Fig. 17.3). (1) Rotate clockwise about O through angle θ to bring OA onto the x-axis (P goes to P_1). (2) Reflect in the x-axis (P_1 to P_2). (3) Rotate anticlockwise by angle θ about O to restore OA to its original position (P_2 to P_3 which is Q).

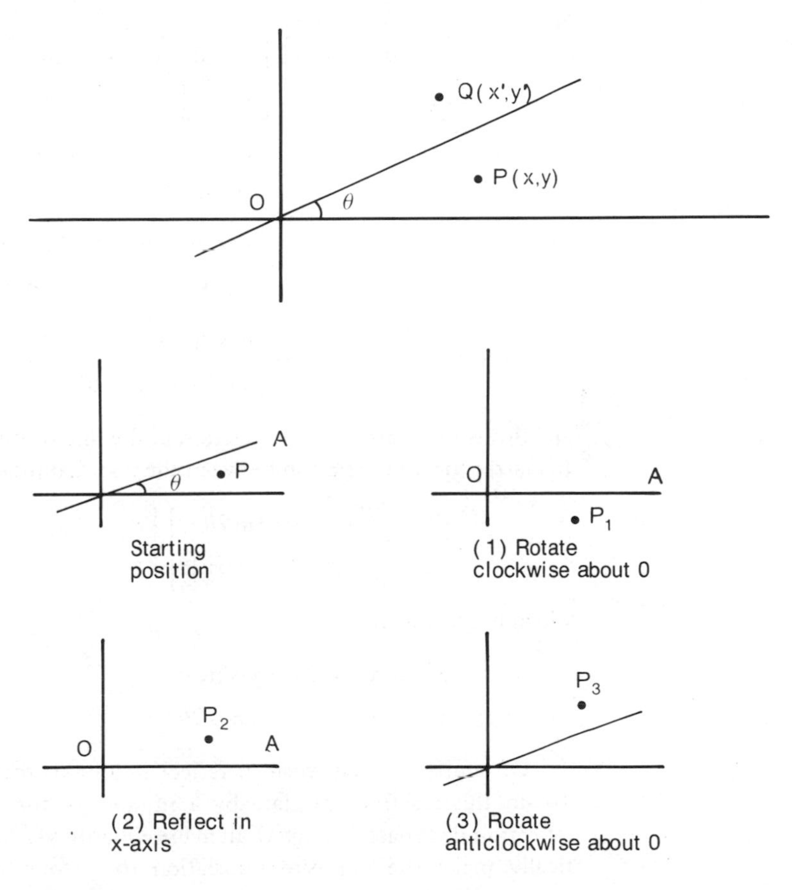

Fig. 17.3

The matrix for a clockwise rotation of θ about O is

$$\begin{bmatrix} \cos\theta & \sin\theta \\ -\sin\theta & \cos\theta \end{bmatrix}$$

for a reflection in the x-axis is

$$\begin{bmatrix} 1 & 0 \\ 0 & -1 \end{bmatrix}$$

and for an anticlockwise rotation about O is

$$\begin{bmatrix} \cos\theta & -\sin\theta \\ \sin\theta & \cos\theta \end{bmatrix}$$

Taking these movements in the appropriate order we have

$$\begin{bmatrix} x' \\ y' \end{bmatrix} = \begin{bmatrix} \cos\theta & -\sin\theta \\ \sin\theta & \cos\theta \end{bmatrix} \begin{bmatrix} 1 & 0 \\ 0 & -1 \end{bmatrix} \begin{bmatrix} \cos\theta & \sin\theta \\ -\sin\theta & \cos\theta \end{bmatrix} \begin{bmatrix} x \\ y \end{bmatrix}$$

Now

$$\begin{aligned} \begin{bmatrix} \cos\theta & -\sin\theta \\ \sin\theta & \cos\theta \end{bmatrix} \begin{bmatrix} 1 & 0 \\ 0 & -1 \end{bmatrix} \begin{bmatrix} \cos\theta & \sin\theta \\ -\sin\theta & \cos\theta \end{bmatrix} &= \begin{bmatrix} \cos\theta & \sin\theta \\ \sin\theta & -\cos\theta \end{bmatrix} \begin{bmatrix} \cos\theta & \sin\theta \\ -\sin\theta & \cos\theta \end{bmatrix} \\ &= \begin{bmatrix} \cos^2\theta - \sin^2\theta & 2\cos\theta\sin\theta \\ 2\cos\theta\sin\theta & -\cos^2\theta + \sin^2\theta \end{bmatrix} \\ &= \begin{bmatrix} \cos 2\theta & \sin 2\theta \\ \sin 2\theta & -\cos 2\theta \end{bmatrix} \end{aligned}$$

and this is the matrix for a reflection in the line through O at angle θ to the x-axis. In matrix form the relation between the coordiantes of P and Q is

$$\begin{bmatrix} x' \\ y' \end{bmatrix} = \begin{bmatrix} \cos 2\theta & \sin 2\theta \\ \sin 2\theta & -\cos 2\theta \end{bmatrix} \begin{bmatrix} x \\ y \end{bmatrix}$$

which is equivalent to

$$\begin{aligned} x' &= x\cos 2\theta + y\sin 2\theta \\ y' &= x\sin 2\theta - y\cos 2\theta \end{aligned} \tag{17.1}$$

Of course we may wish to reflect in an axis that does not pass through O. To do this we first translate by a suitable vector so that the desired axis of reflection does pass through O, then use equations (17.1) to perform the reflection, finally make the opposite translation to restore the axis of reflection to its original position. It is convenient to make the translation parallel to the y-axis and of amount equal to the y-intercept C (see Fig. 17.4). The first translation may be effected by subtracting C from the y-coordinate and the opposite translation by adding C.

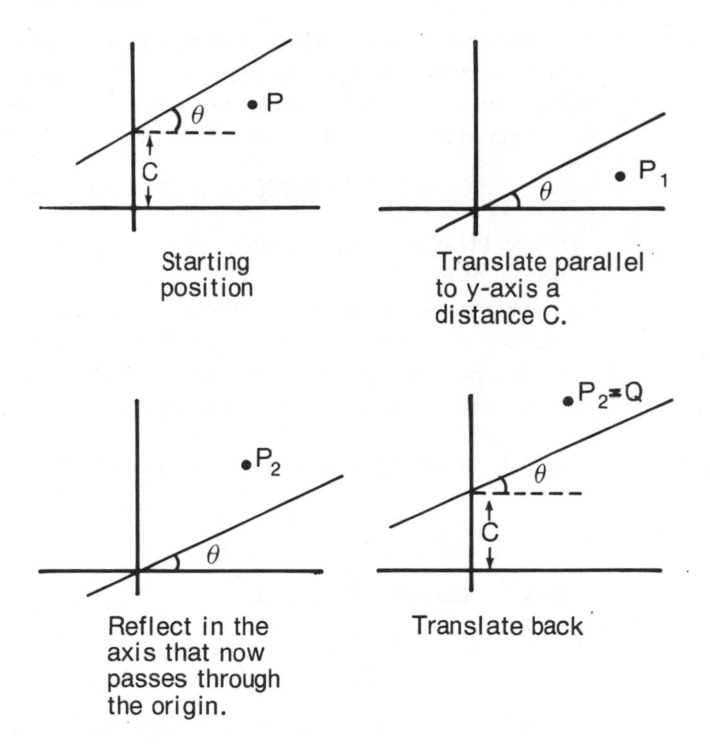

Fig. 17.4

To reflect the point (x,y) in a line at angle θ to the x-axis and making y-intercept C the three steps are applied to the coordinates.

Step 1 $\quad x_1 = x$

$$y_1 = y - C$$

Step 2 (uses equations (17.1))

$$x_2 = x_1 \cos 2\theta + y_1 \sin 2\theta$$

$$y_2 = x_1 \sin 2\theta - y_1 \cos 2\theta$$

Step 3 $\quad x_3 = x_2,$

$$y_3 = y_2 + C.$$

In the computer it is conveneint to store the point in the same stores before and after the reflection. Also steps 2 and 3 may be combined. In Program 17.1 we use XT and YT for x_1 and y_1 so for the point number J of the shape the lines of BASIC that perform the transformation are:

Step 1
```
537  XT = S(1,J) : YT = S(2,J) - C
```

Steps 2 and 3

```
540  S(1,J) = XT*CO + YT*SI
541  S(2,J) = XT*SI - YT*CO + C
```

In elementary mathematics students usually think of angles in degrees and for convenience θ is input in degrees. The computer works with radian measure however, so the angle must be converted. The values of $\cos 2\theta$ and $\sin 2\theta$ may be computed and stored as CO and SI by the line

```
530  CO = COS(2*TH*PI/180) : SI = SIN(2*TH*PI/180)
```

where TH is the angle in degrees. (Note. Some microcomputers use the symbol π instead of PI.)

When the desired axis of reflection is parallel to the y-axis the above approach will not work. This case needs to be dealt with separately. As before the transformation proceeds in three steps but here the algebra is so simple that it reduces to one calculation on the x-coordiante only. The stages are:

Step 1 Translate parallel to the x axis a distance $-D$.

$$x_1 = x - D$$
$$y_1 = y$$

Step 2 Reflect in the y-axis.

$$x_2 = -x_1$$
$$y_2 = y_1$$

Step 3 Translate parallel to the x-axis a distance D.

$$x_3 = x_2 + D$$
$$y_3 = y_2$$

Putting these together we have

$$\begin{aligned} x_3 &= -(x-D)+D \\ &= -x+2D \\ y_3 &= y \end{aligned}$$

Since x_3 and y_3 are stored in the same stores as x and y in the program we only need the one line:

```
565  S(1,J) = -S(1,J) + 2*D
```

which is applied to each point J of the shape.

In lines 500 to 575, Program 17.1 offers the student a reflection and calculates the new coordinates. First in lines 515 to 522 the user indicates if the axis is parallel to the y-axis, this case then being dealt with in lines 550 to 575. For other reflections the y-intercept and angle of the axis are input at line 528. In both cases control passes to line 800 when the calculations are complete for the shape to be drawn in its new position before returning to the menu.

17.6 ROTATION

For a rotation of angle θ with centre at the origin the matrix is

$$\begin{bmatrix} \cos\theta & -\sin\theta \\ \sin\theta & \cos\theta \end{bmatrix}$$

To rotate about a centre that is not at the origin we may proceed in a similar way to that used above for reflection, that is, translate, rotate, then translate back. With the centre of rotation at the point (CX, CY) the equations are:

Step 1
$$x_1 = x - CX$$
$$y_1 = y - CY$$

Step 2
$$x_2 = x_1 \cos\theta - y_1 \sin\theta$$
$$y_2 = x_1 \sin\theta + y_1 \cos\theta$$

Step 3
$$x_3 = x_2 + CX$$
$$y_3 = y_2 + CY$$

As before using CO for $\cos\theta$ and SI for $\sin\theta$ the lines of BASIC for the point J of the shape are:

Step 1

```
630  XT = S(1,J) - CX : YT = S(2,J) - CY
```

Steps 2 and 3

```
635  S(1,J) = XT*CO - YT*SI + CX
636  S(2,J) = XT*SI + YT*CO + CY
```

The rest of the section follows the same pattern as for reflection.

17.7 ENLARGEMENT

The matrix for an enlargement with centre at the origin and scale factor SF is

$$\begin{bmatrix} SF & 0 \\ 0 & SF \end{bmatrix}$$

For a centre other than the origin we may proceed as for reflection or rotation above and translate before and after the enlargement. The equations are:

Step 1
$$x_1 = x - CX$$
$$y_1 = y - CY$$

Step 2
$$x_2 = SF*x_1$$
$$y_2 = SF*y_1$$

Step 3
$$x_3 = x_2 + CX$$
$$y_3 = y_2 + CY$$

and these reduce to

$$x_3 = SF*(x - CX) + CX$$
$$y_3 = SF*(y - CY) + CY$$

and may be performed in one step in the program. For point J the program lines are:

```
725  S(1,J) = SF*(S(1,J) - CX) + CX
726  S(2,J) = SF*(S(2,J) - CY) + CY
```

As before, the shape is drawn in its new position as soon as the calculations are complete.

Program 17.1 – Transformation geometry program

This version is for the BBC computer and some words will need to be changed for other versions of BASIC. The lines that contain these statements are marked with a dagger (†). Line 93 sets the graphics mode and lines 95 and 97 set text and graphics windows. The MOVE and DRAW statements are different in other systems as also is the clear screen CLS.

```
  80  REM**PROGRAM 17.1**
  90  REM**TRANSFORMATION GEOMETRY**
 †93  MODE 4
 †95  VDU28,0,2,39,0
 †97  VDU24,0;0;1279;927
 100  REM**DRAW AXES**
†105  CLS : CLG
 110  XS = 639 : YS = 463 : SQ = 0.9 : SC = 50
†115  MOVE XS,0 : DRAW XS, 927
†120  MOVE 0, YS : DRAW 1279, YS
 125  XL = INT(XS/SC) : YL = INT(YS/(SQ*SC))
 130  FOR J = XS - SC*XL TO XS + SC*XL STEP SC
†135  PLOT 69,J, YS + 4 : PLOT 69,J, YS + 8
 140  NEXT J
 145  FOR K = YS - YL*SC*SQ TO YS + YL*SC*SQ STEP SC*SQ
†150     PLOT 69, XS + 8,K : PLOT 69, XS + 12,K
 155  NEXT K
 200  REM**INPUT SHAPE**
 205  DIM S(2, 11)
†215  CLS
 220  PRINT "PLEASE TYPE COORDINATES OF SHAPE."
 225  PRINT "AND 100,100 TO END."
 230  N = 1
 235  PRINT "POINT"; N;
 240  INPUT S(1,N), S(2,N)
 245  IF S(1,N) >= 100 THEN GOTO 280
 250  IF N > 1 THEN GOTO 260
†255  MOVE XS + SC*S(1, 1), YS + SQ*SC*S(2,1) : GOTO 265
```

```
†260  DRAW XS + SC*S(1,N), YS + SQ*SC*S(2,N)
 265  N = N + 1 : GOTO 235
 280  S(1,N) = S(1,1) : S(2,N) = S(2,1)
†285  DRAW XS + SC*S(1,N), YS + SQ*SC*S(2,N)
 300  REM**OFFER MENU**
†305  CLS
 310  PRINT "TRANSLATE(1), REFECT(2), ROTATE(3),"
 311  PRINT "ENLARGE(4) OR END(5) –"
 312  PRINT "WHICH NUMBER";
 315  INPUT CH
 320  IF CH < 1 OR CH > 5 THEN GOTO 305
 325  ON CH GOTO 400, 500, 600, 700, 900
 400  REM**TRANSLATE**
†410  CLS
 415  PRINT "TRANSLATION – WHAT X, Y COMPONENTS"
 420  INPUT TX, TY
 425  FOR J = 1 TO N
 430     S(1,J) = S(1,J) + TX
 431     S(2,J) = S(2,J) + TY
 435  NEXT J
 440  GOTO 800
 500  REM**REFLECTION**
†510  CLS
 515  PRINT "REFLECTION – IS AXIS PARALLEL TO"
 516  PRINT "Y-AXIS? (Y OR N)";
 520  INPUT R$
 521  IF R$ = "Y" THEN GOTO 550
 522  IF R$ <> "N" THEN GOTO 510
†525  CLS
 526  PRINT "WHAT AXIS OF REFLECTION?"
 527  PRINT "TYPE Y-INTERCEPT, ANGLE";
 528  INPUT C, TH
 530  CO = COS(2*TH*PI/180) : SI = SIN(2*TH*PI/180)
 535  FOR J = 1 TO N
 537     XT = S(1,J) : YT = S(2,J) - C
 540     S(1,J) = XT*CO + YT*SI
 541     S(2,J) = XT*SI - YT*CO + C
 545  NEXT J
 547  GOTO 800
†550  CLS
 551  PRINT "REFLECTION PARALLEL TO Y-AXIS."
 552  PRINT "WHAT X-INTERCEPT";
 555  INPUT D
 560  FOR J = 1 TO N
 565     S(1,J) = -S(1,J) + 2*D
 570  NEXT J
 575  GOTO 800
```

```
 600  REM**ROTATION**
†605  CLS
 610  PRINT "ROTATION – WHAT CENTRE-X, CENTRE-Y,"
 611  PRINT "ANGLE";
 615  INPUT CX, CY, TH
 620  CO = COS(TH*PI/180) : SI = SIN(TH*PI/180)
 625  FOR J = 1 TO N
 630    XT = S(1,J) - CX : YT = S(2,J) - CY
 635    S(1,J) = XT*CO - YT*SI + CX
 636    S(2,J) = XT*SI + YT*CO + CY
 640  NEXT J
 645  GOTO 800
 700  REM**ENLARGE**
†705  CLS
 710  PRINT "ENLARGEMENT – WHAT CENTRE-X, CENTRE-Y,"
 711  PRINT "SCALE FACTOR";
 715  INPUT CX, CY, SF
 720  FOR J = 1 TO N
 725    S(1,J) = (S(1,J) - CX)*SF + CX
 726    S(2,J) = (S(2,J) - CY)*SF + CY
 730  NEXT J
 735  GOTO 800
 800  REM**DRAW SHAPE**
†810  MOVE XS + SC*S(1,1), YS + SQ*SC*S(2,1)
 815  FOR J = 2 TO N
†820    DRAW XS + SC*S(1,J), YS + SQ*SC*S(2,J)
 825  NEXT J
 830  GOTO 300
 900  REM**END PROGRAM**
†910  CLS
 915  PRINT "PROGRAM FINISHED"
 999  END
```

One way of using Program 17.1 is for a group or class to work together. After the initial shape has been designed and drawn various transformations can be described and the class asked where the image will appear. When everyone has made their suggestions the computer can be fed with the relevant information and will show how close the actual image is to the expected position. Alternatively it can be stated where the image is to appear and the class asked to suggest transformations that can bring this about. We believe that such activity greatly strengthens the students' ability to visualise various transformations. If the reader has no experience of this sort of interactive work with transformations we suggest that he tries the program himself. Most people find the judgements involved more difficult than they expect, and this emphasises how little experience of transformations many of us have. Static diagrams in books do not give the same opportunity to build mental images as does a dynamic computer program.

17.8 PRESENTING THE IMAGE THEN ESTIMATING THE TRANSFORMATION

A development of the program can be made in which the computer draws a shape and its image under some randomly chosen transformation. The student then has to try to judge what transformation has taken place. The appropriate parameters are then chosen and the computer draws the student's image. For this purpose a shape that clearly shows its position is required. Clearly any symmetry in the shape could lead to confusion. We have used a 'flag' that is easily defined in a coordinate system (see Fig. 17.5) the coordinate matrix S being

$$\begin{matrix} 1 & 1 & 2 & 1 \\ 1 & 3 & 2 & 2 \end{matrix}$$

We apply a simple transformation to S to produce a starting position and the flag is drawn in this position. A record is kept of the starting position.

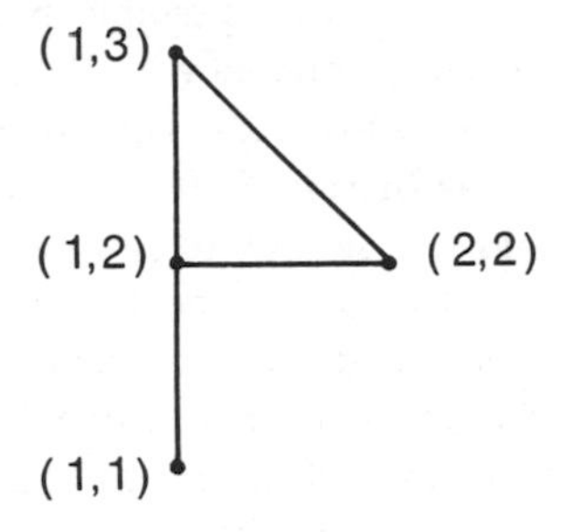

Fig. 17.5

A transformation is now applied to this initial shape and its image drawn on the screen, preferably in a different colour. The student is then invited to judge what transformation has been applied and the student's image is drawn. The objective is to try to get the student's image on the computer's image. This activity is also rather more difficult than it sounds and provides valuable experience in visualising transformations.

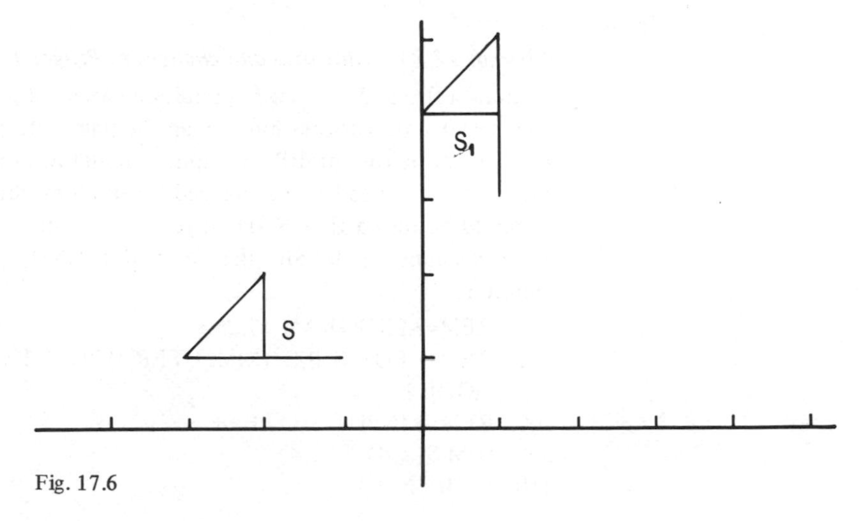

Fig. 17.6

For example, suppose the computer shows the starting position S and the image S_1 shown in Fig. 17.6. The student decides that this is a reflection and has to decide the y-intercept and the angle of the axis (that is, 2,135). How well he judges this is apparent from the resulting image that is drawn.

For a program to do this, much of the material in Program 17.1 may be used. It is preferable to put the sections to effect the transformations into subroutines so each section will end with RETURN in place of GOTO 800. The section from line 200 to 285 will need to be rewritten so that the computer sets the coordinates of the flag and chooses a transformation randomly from the list. Random numbers are also used to choose the parameters of the transformation and the flag is drawn. The coordinates of this initial flag are saved, then it is transformed again by a second randomly chosen transformation. The flag is drawn again in its new position and this forms the target for the student. Movements are offered just as in Program 17.1.

The list of transformations to be used may be varied as also may the ranges of the parameters, but we give one example based on these ideas as Program 17.2. The chief change when compared with Program 17.1 is to lines 200 to 296. A shape is defined in lines 200 to 220. The random number CH chosen in line 226 determines which type of transformation to apply to the shape, and from line 228 control branches to a section to choose the parameters. For example, if CH is 2 then reflection is selected and line 272 chooses the parameters randomly. The shape is then transformed by the section of the reflection subroutine starting at line 530 and control returns to line 230. The shape is drawn by the subroutine in line 800 and control returns to 232. Lines 232 to 238 keep a copy of the shape in array T. A new position for the flag is determined similarly in lines 240 to 244 and it, too, is drawn. The array S is then reset to the first flag in lines 250 to 256 and control passes to the original program at line 300. The student is thus offered choices as before and can attempt to map the first flag (white) onto the second (yellow).

By means of programs of this sort students may be able to gain practice in visualising transformations that is difficult to obtain in other ways. Many variations of the basic program are possible to vary the difficulty and options offered.

Program 17.2 – Additions and changes to Program 17.1

To draw a flag and its image in randomly selected positions. Students then select a transformation to map the flag to its image if they can! This version is for the BBC computer model B. Lines marked with a dagger (†) may need to be changed for other machines. In particular it should be noted that RND(4) returns an integer 1, 2, 3 or 4. For other versions of BASIC the form INT(RND(6)*4 + 1) may be required.

```
 80  REM**PROGRAM 17.2**
 81  REM**FOR B.B.C. COMPUTER MODEL B**
†93  MODE 1
200  REM**INPUT SHAPE**
205  DIM S(2,4),T(2,4)
210  F=0 : N=4
```

```
 215  FOR J = 1 TO 4
 217     READ S(1,J),S(2,J)
 219  NEXT J
 220  DATA 1,1,1,3,2,2,1,2
 224  REM**CHOOSE INITIAL POSITION**
†226  CH = RND(4)
 228  ON CH GOSUB 260, 270, 280, 290
 230  GOSUB 800
 232  REM**COPY SHAPE**
 234  FOR J = 1 TO 4
 236     T(1,J) = S(1,J) : T(2,J) = S(2,J)
 238  NEXT J
 240  REM**NEW POSITION FOR SHAPE**
†242  CH = RND(4)
 244  ON CH GOSUB 260, 270, 280, 290
†245  GCOL 0,2
 246  GOSUB 800
†247  GCOL 0,3
 250  REM**RESTORE ORIGINAL SHAPE**
 252  FOR J = 1 TO 4
 254     S(1,J) = T(1,J) : S(2,J) = T(2,J)
 256  NEXT J
 258  GOTO 300
 260  REM**TRANSLATION PARAMETERS**
†262  TX = RND(9) - 5 : TY = RND(9) - 5
 264  GOSUB 425
 266  RETURN
 270  REM**REFLECTION PARAMETERS**
†272  C = RND(7) - 4 : TH = RND(8)*45
 274  GOSUB 530
 276  RETURN
 280  REM**ROTATION PARAMETERS**
†282  CX = RND(7) - 4 : CY = RND(7) - 4 : TH = RND(8)*45
 284  GOSUB 620
 286  RETURN
 325  ON CH GOSUB 400, 500, 600, 700, 900
 330  IF F = 1 THEN END
 335  GOSUB 800
 340  GOTO 300
 440  RETURN
 547  RETURN
 575  RETURN
 645  RETURN
 735  RETURN
 830  RETURN
 920  F = 1
 925  RETURN
```

18

Writing mathematical software

The writing of elaborate software that can cope with any conceivable case or user is a time-consuming business that is best left to those with the necessary expertise and time. In spite of this, we believe that teachers will sometimes wish to construct their own programs to meet a specific need for which ready made material is not available. Many of the techniques that can be used are illustrated elsewhere in this book in the various applications of microcomputers that are suggested there. The speed with which a program can be written, and its convenience and trustworthiness in use, can be greatly affected by the adoption of a sound approach and a good programming style. Here we shall trace through the development from the initial idea to a workable program as a means of illustrating many of the principles involved. We shall keep to relatively simple structures and commonly used elements of BASIC in the belief that this is what many teachers, who have but an elementary knowledge of programming, need. Often there are more elegant or efficient ways of achieving the same result, but we are chiefly concerned with clarity and ease of obtaining a result that works. We have tried, as far as possible, to stick to those methods that can be used on any microcomputer with little, if any, modification. It is for this reason that we use such things as PRINT statements to produce blank lines on the VDU rather than resorting to the various ways of accessing directly a desired position on the screen.

18.1 THE INITIAL IDEA

The start of program development is the initial idea, probably rather vague at first. In our example we use the idea that children (and adults for that matter) often have difficulty in making an estimate of the value of an arithmetic expression. Teachers find it difficult to give enough experience of this, especially to pupils individually, so why not use a computer to present suitable expressions? The student can then make an estimate and the computer a response that depends on how good the estimate is. This, then, is our starting point, but how do we turn it into a working program?

18.2 PROBLEM ANALYSIS

Having satisfied ourselves that the idea is worth the time needed to develop it, we next do a complete analysis of the problem. This is one of the most important parts of the process. The aim is to decide exactly what we want the computer to do and describe it as a series of specific tasks. These are then put into a logical order showing which ones lead to which others.

For the estimating task we decide to arrange for a variety of types of expression to be available and for the student to be able to choose the type. A later version may allow the computer to mix types of question but we will not include that for the moment. Numbers for the expression will be generated randomly, but the range will be within chosen limits. The student's response to the expression requires various levels of reaction by the computer. 'Good' for a close estimate, encouragement for a reasonable one, and some help and exhortation to try again if it is poor, are the minimum requirements. It would be helpful to the student to know how close he is getting to the exact value. He could be told his percentage error and shown the true value. In cases where he has no idea of the value the option to see it could be offered. We also need to decide whether to present a fixed number of questions or let the student decide whether or not to have another.

Now that we have a fairly clear idea of what the program is to do we can list the tasks that have to be done as follows:

- Title page and description of the task.
- Offer the types of question (format) and student chooses one.
- Generate random numbers for the expression.
- Display the expression.
- Calculate correct value.
- Student inputs estimate of value.
- Check student's estimate.
- Reward a good estimate.
- Encourage a fair estimate and offer a second guess.
- For a poor estimate offer a second attempt or to see exact value.
- Offer a further question.
- Offer a change of format.

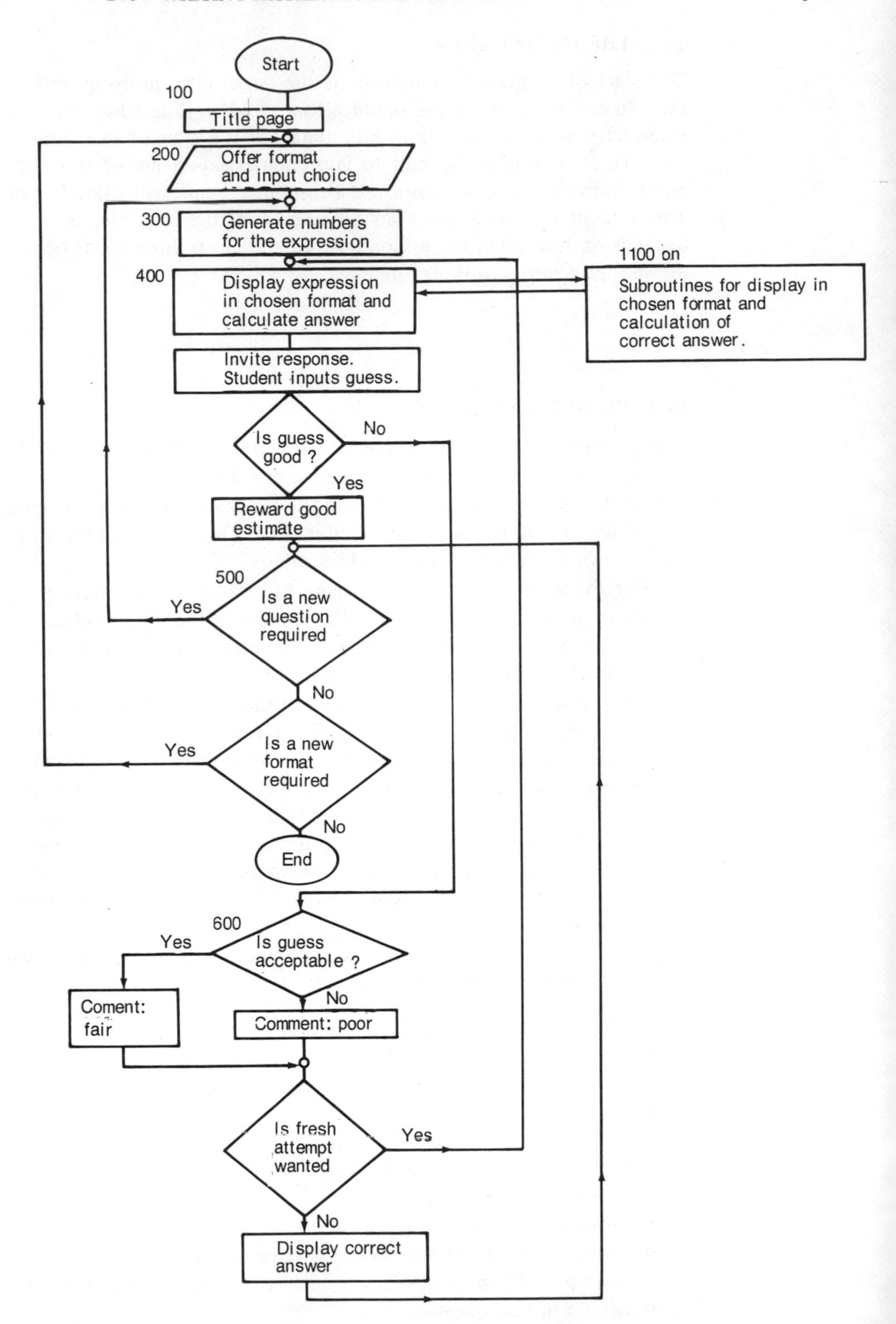

Fig. 18.1 – Flowchart for the estimation problem.

It is now necessary to put these tasks into a correct logical order and show the route the computer is to take through them. The classical way to do this is by means of a flowchart. Some programmers claim that they do not need a flowchart, but they still have to work out the overall structure of the program. In our view the discipline of constructing a good flowchart, or structure diagram, leads to a more logical program that is much easier to understand and modify. The search for programming or typing errors is also made easier. It is very frustrating to write a program in an hour or so then have to spend another six or seven hours eliminating the mistakes. Far better to take a little more trouble early on and to structure it well. However, there is the other danger of going to extremes in putting every tiny step of the process in its own flowchart box. The result is difficult to follow and not much help in seeing just how each section fits into the whole. In Fig. 18.1 we give a possible flowchart for the estimation program. Each box contains one task and will correspond to a section of the final program. The allocation of line numbers has also been planned at this stage, and the line numbers for the start of tasks have been written against the boxes at key points in the program. Now we are ready to write the program, one section at a time, in BASIC.

18.3 WRITING THE PROGRAM

It may seem logical to start the program writing at the beginning; however, this is often not the best way to do it. We recommend that the writing proceeds one section at a time, and that each is tested before attempting the next. The finding and elimination of errors is much easier if approached in this way. We will start with the heart of the program in lines 400 to 500. Other sections can be replaced by dummy lines so that this first section may be tested, these dummy lines being removed later as the other sections are written. The store CH will be used for the number of the chosen format. Eventually there will be six of these and CH can take any value from 1 to 6, but for now we will set CH = 1 and write just one format subroutine starting at line 1100. The expressions will involve no more than three numbers and P, Q, R will be used. Later we will write the routine to select these, but, for now, set P = 12.5, Q = 47 and R = 8.3. We thus include the dummy lines:

```
200  CH = 1
300  P = 12.5 : Q = 47 : R = 8.3
600  PRINT "GUESS NOT GOOD." : END
```

Expression number 1 will be P × Q × R and the subroutine starting at line 1100 displays this on the screen. Not all computers have a '×' sign in the character set and we may have to use a letter 'X'. Line 1110 is included to evaluate the answer AN. The program lines are:

```
1100  REM**FORMAT P×Q×R**
1110  AN = P*Q*R
1120  PRINT "        ";P;" × ";Q;" × ";R
1125  RETURN
```

We are now ready to write the section starting at line 400.

To display in the chosen format we may use the statement

```
410  ON CH GOSUB 1100,1200,1300,1400,1500,1600
```

so that the value of CH determines which of the six subroutines will be called. For now CH = 1, but later when CH can have any of the values 1 to 6 we must remember to add the other five subroutines. The student's guess will be stored in G by the line

```
435  INPUT G
```

and then a test for a good guess must be devised. Suppose we choose to call any guess within 10% of the correct answer a good one. We then calculate the absolute percentage error F and test it with the lines

```
440  F = ABS(AN - G)/AN*100
445  IF F>10 THEN GOTO 600
```

Since we intend to tell the student the percentage error we do not want too many decimal places in F. One decimal place should be enough so we include the line

```
442  F = INT(F*10 + 0.5)/10
```

to round F to one decimal place. The rest of the section is straightforward and at this stage in its development the program is:

```
 200  CH = 1
 300  P = 12.5 : Q = 47 : R = 8.3
 400  REM**DISPLAY QUESTION**
 405  CLS : PRINT : PRINT : PRINT
 410  ON CH GOSUB 1100, 1200, 1300, 1400, 1500, 1600
 420  REM** INPUT GUESS AND TEST**
 425  PRINT : PRINT
 430  PRINT "WHAT IS YOUR GUESS";
 435  INPUT G
 440  F = ABS(AN - G)/AN*100
 442  F = INT(F*10 + 0.5)/10
 445  IF F>10 THEN GOTO 600
 450  PRINT : PRINT
 455  PRINT "        WELL DONE"
 460  PRINT : PRINT "THE CORRECT ANSWER IS "; AN
 462  PRINT
 465  PRINT "SO YOU WERE ONLY ";F;"% OUT."
 599  END
 600  PRINT "GUESS NOT GOOD" : END
1100  REM**FORMAT P×Q×R SUBROUTINE**
1110  AN = P*Q*R
1120  PRINT "        ";P;" × ";Q;" × ";R
1125  RETURN
```

We have used extra PRINT statements to produce black lines on the VDU screen since all versions of BASIC accept this. Individual machines may have better ways of achieving the required layout and the reader will no doubt be able to use these as desired. The CLS instruction to clear the screen differs on many machines and will quite likely have to be changed. The program may now be tested both with good and bad guesses to check that the appropriate response occurs.

18.4 RANDOM NUMBER GENERATION

The next section to be written could be the generation of the random numbers P, Q and R. Microcomputers have a way of generating random numbers and we will assume, as is often the case, that RND(1) produces a random number in the range 0 to 1. The precise form of this statement may be rather different on some machines. This range is too small for our purpose and we decide to have P in the range 0 to 100 with one decimal place. RND(1)*1000 will produce a random number in the range 0 to 1000 so by taking the integer part then dividing by 10 we get a number in the required range. The program line is thus:

```
310  P = INT(RND(1)*1000)/10
```

For Q in the range 0 to 50 with two decimal places and R in the same range but one decimal place we use:

```
320  Q = INT(RND(1)*5000)/100
330  R = INT(RND(1)*500)/10
```

If different ranges or different numbers of decimal places are desired then these lines may be varied accordingly. The reader may like to experiment with various values. The program may now be tested by running several times to see that satisfactory numbers are generated.

18.5 DEALING WITH A POOR ESTIMATE

A guess that is more than 10% wrong causes control to transfer to line 600. In the section that starts there fair estimates must be distinguished from bad ones. A fair estimate, we will say, is one that is between 10% and 25% from the correct value. The test in line 445 ensures that on arriving at line 600 we have $F > 10$ so we only need

```
625  IF F > 25 THEN GOTO 690
```

to separate bad estimates from the rest. These will be dealt with at 690 before returning to line 630. Having told the student that the estimate is fair or poor, as the case may be, we will now offer the option of another guess. He must answer 'yes' or 'no' to the question 'Do you want another guess?' and this could be done with an INPUT statement. However, for convenience a single key will be all that is needed to trigger the desired response from the computer. In BASIC this is achieved by making use of GET; however, the use of this varies somewhat from machine to machine. Sometimes the instruction is GET R$ but for our example the version is R$ = GET$. The result is that R$ will contain a single character according to which key was pressed. Some computers wait for a key to be pressed but others do not. Of course most of the time no key is being pressed so R$ contains no character at all (a null string). The technique is to make the computer repeatedly go back to GET until a character is found. This may be done by a conditional statement that branches to itself. In our case we use

```
635 R$ = GET$ : IF R$ = "" THEN GOTO 635
```

Note that in the test there is no character at all between the quotation marks, not even a space. Thus R$ is compared with a completely empty string (the null string). For machines like the BBC computer where GET waits for a key to be pressed the conditional in line 635 is not required; however, it does no harm, though it is not used in that case. We want the student to press Y for yes or N for no in answer to the question. It is wise to arrange for any other character to be rejected so that an accidental touching of a key cannot produce an unwanted result. First we test for a 'yes' answer with

```
640 IF R$ = "Y" THEN GOTO 400
```

then reject anything other than a 'no' with

```
645 IF R$ <> "N" THEN GOTO 635
```

If the response has been 'N' for no then the exact answer can be revealed to the student. The section so far now reads as follows:

```
600 REM**GUESS NOT GOOD**
610 REM**IS GUESS ACCEPTABLE?**
615 PRINT : PRINT
620 PRINT "YOUR ANSWER IS"; F;"% OUT."
622 PRINT
625 IF F > 25 THEN GOTO 690
626 PRINT "THIS IS A USEFUL ESTIMATE, BUT ..."
627 PRINT
630 PRINT "WOULD YOU LIKE TO GUESS AGAIN (Y OR N)";
635 R$ = GET$ : IF R$ = "" THEN GOTO 635
640 IF R$ = "Y" THEN GOTO 400
645 IF R$ <> "N" THEN GOTO 635
```

Lines will be inserted here to display the correct answer – see below.

```
690 REM**POOR GUESS**
694 PRINT "THIS IS NOT A GOOD ESTIMATE."
696 PRINT
698 GOTO 630
```

If the correct answer is to be displayed it would be useful for the student to see the question immediately above it. We will therefore need to display the expression again and, since CH still contains the number of the chosen format, we may again use it to call the appropriate subroutine. Naturally by this time the screen is getting rather cluttered, so first we will clear it. The program lines are:

```
650 REM**DISPLAY CORRECT ANSWER**
655 CLS
660 PRINT : PRINT : PRINT
665 PRINT "THE CORRECT ANSWER TO"
670 PRINT : PRINT
675 ON CH GOSUB 1100,1200,1300,1400,1500,1600
680 PRINT : PRINT
682 PRINT "       IS "; AN
685 GOTO 500
```

Line 500 has not yet been entered and is where we plan to offer a new question. To enable this section to be tested we will include the dummy line

```
500 PRINT "OFFER NEW QUESTION" : END
```

18.6 MUNU TO OFFER CHOICE OF EXPRESSION

Having made the main parts of the program function correctly, it is now the time to tackle the problem of offering a choice of format. We will use this to illustrate the technique of menu selection. A list of possible options is given together with a symbol for choosing each. Letters, numbers, or any other character may be used and the GET statement, as described above, can be used to collect the choice. We will offer six possible types of expression and number them 1 to 6, the chosen number being put into CH by the lines:

```
245 CH$ = GET$ : IF CH$ = "" THEN GOTO 245
247 CH = VAL(CH$)
250 IF CH<1 OR CH>6 THEN GOTO 245
```

It may be wondered why we do not use CH = GET directly. The reason is that on some machines it is the code for the character that is produced by GET rather than its value. The code for the key marked '1' is usually 49. By inputting the character into CH$ then using the function VAL to obtain its value we ensure that the desired number is returned in CH whatever system the machine uses. The reader may be able to simplify this section for his particular machine.

Many formats are possible but for our example we will use the six given below.

(1) $P \times Q \times R$ (2) $(P+Q) \times R$

(3) $\dfrac{P}{Q}$ (4) $P+(Q \times R)$

(5) $\dfrac{P \times Q}{R}$ (6) $\dfrac{P+Q}{R}$

This is also how the menu will appear on the screen together with the invitation to select one of the numbers 1–6. The program lines to display the menu are straightforward and the complete section is to be found in the full listing of the program at the end of the chapter as lines 200 to 250. To test that the section is working correctly we cannot at the moment use the whole program so far, because only the first of the subroutines selected in line 410 by the value of CH has been written. A way out is to put in some extra lines to test this part alone without falling into the rest of the program. These must be removed as soon as we are satisfied that the menu works correctly. A possible way to do this is with:

```
260  PRINT "CHOSEN FORMAT IS NO. "; CH
265  END
```

Once the menu is working correctly it is a good time to enter the remaining five subroutines to display the expression in formats 2 to 6. These are straightforward pieces of programming to print the numbers together with the operation symbols, fraction lines etc. Our examples will be found in lines 1200 to 1630 of the full listing. These subroutines ought also to be tested on their own before we couple them with the rest of the program. We want to give CH a value then apply the appropriate subroutine to print dummy values of P, Q and R in the chosen format. Again we use some program lines that are removed as soon as the test is complete. We suggest something such as:

```
100  INPUT CH
110  P = 12.5 : Q = 47 : R = 8.3
120  GOTO 400
415  END
```

This may be run six times, inputting the numbers 1 to 6 in turn so that all six subroutines are tested. We may then remove the extra lines and test the program as a whole.

All that now remains to be done is to write the introduction that in our example will be found in lines 100 to 130, and the section to offer a new format, lines 550 to 580.

18.7 USING THE PROGRAM

The final task is to try out the program and see if it performs as intended. It may be varied in several ways to meet a variety of circumstances, but probably the generation of values for P, Q and R is the part most likely to need changing. We may decide to allow different ranges or numbers of decimal places. For some pupils we may prefer to use integers. Different sections for the generation of P, Q and R may be selected by the student choosing a level of difficulty. We have found this a viable way of adding flexibility to the program. Three different sections for the generation of P, Q and R are accessed by the value of a 'difficulty' number 1, 2 or 3 selected by the student. The available formats could also be changed and this would entail altering the menu and adjusting the format subroutines.

We have found that children can usefully use the program at a much younger age than we had expected. Subjective assessment of children working with it suggests to us that their abaility to estimate in this context improves considerably with use of such a program. Clearly the extent to which these skills can then be applied generally needs further investigation, but we believe that a program of this sort may fill a need in the mathematics classroom, and we invite the reader to experiment and develop the idea.

18.8 PROGRAM WRITING IN GENERAL

The approach used in this example program, and many of the techniques, will be found applicable to a variety of uses of a micrcocomputer in mathematics teaching and learning. We believe that even for the teacher with a modest amount of programming experience the adoption of a systematic approach makes for a greatly improved result. For the advanced programmer such methods are part of his stock in trade.

Another great advantage of structuring a program in this way is that it makes transferral to another machine much more manageable. It is often wise to bear in mind from the outset that the program should be easily 'portable' between microcomputers and thus to put all machine specific commands, for example, for graphics, in separate subroutines.

It is possible to go much further than we have done here in using a modular structure for a program and this makes for good programming practice. Some computers, such as the BBC allow the definition of procedures and sometimes the whole of the work in a program can be done in these procedures. The main program then consists of a series of calls to the procedures as required and thus manages the task. Choosing names for the procedures that describe what they do results in a main program that is very readable. Interested readers who wish to improve their programming style are recommended to consult one of the many books on structured programming.

For such a program to be usable by other teachers it will be necessary to produce adequate documentation explaining how to use the program.

Program 18.1 – The full estimation program

Lines marked with a dagger (†) contain words that may need to be changed for some microcomputers. This version works on the BBC microcomputer as it stands.

```
  90  REM**PROGRAM 18.1**
 100  REM**GUESTIMATE**
†110  CLS
 115  PRINT : PRINT : PRINT : PRINT "        GUESTIMATE"
 116  PRINT "        ————————————"
 120  PRINT : PRINT : PRINT "TO HELP YOU TO ESTIMATE
      THE VALUE OF"
 121  PRINT "EXPRESSIONS IN ARITHMETIC."
 125  PRINT : PRINT "PRESS ANY KEY TO CONTINUE"
†130  C$ = GET$ : IF C$ = "" THEN GOTO 130
```

```
 200 REM**CHOICE OF FORMAT**
†210 CLS
 215 PRINT : PRINT : PRINT "THE COMPUTER WILL SHOW
     SOME EXPRESSIONS"
 216 PRINT "USING NUMBERS."
 220 PRINT : PRINT "YOU THEN SEE HOW WELL YOU CAN
     ESTIMATE"
 221 PRINT "THE ANSWER TO THE EXPRESSION."
 225 PRINT : PRINT "PLEASE CHOOSE THE TYPE OF
     EXPRESSION"
 226 PRINT "YOU WANT. TYPE THE NUMBER."
 230 PRINT : PRINT
 231 PRINT "   (1) P X Q X R   (2) (P + Q) X R"
 235 PRINT : PRINT
 236 PRINT "          P"
 237 PRINT "   (3)  ---              (4)  P + (Q X R)"
 238 PRINT "          Q"
 240 PRINT : PRINT
 241 PRINT "         P X Q              P + Q"
 242 PRINT "   (5)  -------        (6)  -------"
 243 PRINT "           R                  R"
†245 CH$ = GET$ : IF CH$ = "" THEN GOTO 245
 247 CH = VAL(CH$)
 250 IF CH < 1 OR CH > 6 THEN GOTO 245
 300 REM**GENERATE NUMBERS**
†310 P = INT(RND(1)*1000)/10
†320 Q = INT(RND(1)*5000)/100
†330 R = INT(RND(1)*500)/10
 400 REM**DISPLAY QUESTION**
†405 CLS : PRINT : PRINT : PRINT
 410 ON CH GOSUB 1100, 1200, 1300, 1400, 1500, 1600
 420 REM**INPUT GUESS AND TEST**
 425 PRINT : PRINT
 430 PRINT "WHAT IS YOUR GUESS";
 435 INPUT G
 440 F = ABS(AN - G)/AN*100
 442 F = INT(F*10 + 0.5)/10
 443 AN = INT(AN*100 + 0.5)/100
 445 IF F > 10 THEN GOTO 600
 450 PRINT : PRINT
 455 PRINT "        WELL DONE"
 460 PRINT : PRINT "THE CORRECT ANSWER IS "; AN
 462 PRINT
 465 PRINT "SO YOU WERE ONLY "; F; "% OUT."
 500 REM**NEW QUESTION REQUIRED?**
 510 PRINT : PRINT "WOULD YOU LIKE ANOTHER"
 511 PRINT "QUESTION (Y OR N)?";
```

```
†515  R$ = GET$ : IF R$ = "" THEN GOTO 515
 520  IF R$ = "Y" THEN GOTO 300
 525  IF R$ <> "N" THEN GOTO 515
 550  REM**OFFER CHANGE OF FORMAT**
†555  CLS
 560  PRINT : PRINT : PRINT
 565  PRINT "WOULD YOU LIKE A DIFFERENT KIND OF"
 566  PRINT "QUESTION (Y OR N)?";
†570  R$ = GET$ : IF R$ = "" THEN GOTO 570
 575  IF R$ = "Y" THEN GOTO 200
 580  IF R$ <> "N" THEN GOTO 570
†585  CLS
 590  PRINT : PRINT : PRINT "        PROGRAM FINISHED."
 599  END
 600  REM**GUESS NOT GOOD**
 610  REM**IS GUESS ACCEPTABLE**
 615  PRINT : PRINT
 620  PRINT "YOUR ANSWER IS "; F; "% OUT."
 622  PRINT
 625  IF F > 25 THEN GOTO 690
 626  PRINT "THIS IS STILL A USEFUL ESTIMATE, BUT ..."
 627  PRINT
 630  PRINT "WOULD YOU LIKE TO GUESS AGAIN (Y OR N)?";
†635  R$ = GET$ : IF R$ = "" THEN GOTO 635
 640  IF R$ = "Y" THEN GOTO 400
 645  IF R$ <> "N" THEN GOTO 635
 650  REM**DISPLAY CORRECT ANSWER**
†655  CLS
 660  PRINT "THE CORRECT ANSWER TO"
 670  PRINT : PRINT
 675  ON CH GOSUB 1100, 1200, 1300, 1400, 1500, 1600
 680  PRINT : PRINT
 682  PRINT "        IS   "; AN
 685  GOTO 500
 690  REM**POOR GUESS**
 694  PRINT "THIS IS NOT A GOOD ESTIMATE."
 696  PRINT
 698  GOTO 630
1100  REM**FORMAT P X Q X R**
1110  AN = P*Q*R
1120  PRINT "        "; P; " X "; Q; " X "; R
1125  RETURN
1200  REM**FORMAT (P + Q) X R**
1210  AN = (P + Q)*R
1220  PRINT "        ("; P;" + "; Q;") X "; R
1225  RETURN
1300  REM**FORMAT P/Q**
```

```
1310 AN = P/Q
1320 PRINT "         ";P
1321 PRINT "        ——"
1322 PRINT "         ";Q
1330 RETURN
1400 REM**FORMAT P + (Q X R)**
1410 AN = P + (Q*R)
1420 PRINT "         "; P; " + ("; Q;" X "; R;")"
1430 RETURN
1500 REM**FORMAT (P X Q)/R**
1510 AN = (P*Q)/R
1520 PRINT "           "; P; " X "; Q
1521 PRINT "          ————————————"
1522 PRINT "                  "; R
1530 RETURN
1600 REM**FORMAT (P + Q)/R**
1610 AN = (P + Q)/R
1620 PRINT "           "; P; " + "; Q
1621 PRINT "          ————————————"
1622 PRINT "                  "; R
1630 RETURN
```

19

Using computers in schools

We have tried to show ways in which a microcomputer may be fruitfully used to help in the learning and teaching of many areas of school mathematics. We have *not* adopted the approach: 'so you have a microcomputer, what kind of mathematics can you do with it?', but rather: 'what kind of mathematics can you do with a microcomputer that you could not readily do *any* other way?' However, if there is a case for the use of microcomputers in mathematics teaching there remain many problems concerned with its management within the school's mathematics department that need to be considered. In this section we try to raise the main issues that have come our way – not because we feel there are any obvious or easy answers but because we feel that if a whole department is to become involved it must address itself to these problems and evolve its own style. Only by the development of such departmental approaches will the use of microcomputers in mathematics teaching reach its full potential and have a significant effect on the curriculum.

Before considering some of the problems it may help to identify some of the approaches that have been taken so far in the use of microcomputers in mathematics.

19.1 SOME CURRENT APPROACHES

19.1.1 Individual learning

Here a specific skill, such as converting numbers in base 10 to binary, could be taught by using a program that produces frames of text which explain the process.

Movement from one frame to the next can be controlled by assessing the student's response to one or more questions about the text. This kind of programmed learning can be particularly suitable when individual pupils have been absent during class lessons in which the skill was introduced. There are specialised computer languages, called *author languages* (such as PILOT), that have been designed to make the production of such training programs more simple.

19.1.2 Testing and improving skills

The computer's ability to generate random questions and to measure the time of responses is easily applied to skill testing. The way the pupil interacts with the program may be quite varied. In a typical drill and practice program on multiplication tables the 'level' of difficulty and the time limit for each question may be set by teacher or pupil. The program produces a set of random questions and keeps track of the number of correct responses. In another on equivalent fractions it may just be used to generate a quick 'mental arithmetic' test for the whole class without itself doing any checking on the responses. Other styles include the use of games where two pupils (or groups of pupils) compete in a game against each other where points are scored or moves made by answering computer-generated questions on the specific skill.

19.1.3 Specific teaching aids

Just as a teacher may choose to use an OHP, a video recording or some slides to enhance his teaching he may choose to use a microcomputer running a suitable program. The style of use may vary considerably. For example, there are programs to help teach multiplication of matrices that function as an animated blackboard – highlighting the relevant parts of the process by making the corresponding numbers flash on the screen. A program to help teach rotation may enable the class to define an object on a co-ordinate grid and to specify a centre and angle of rotation, leaving the computer to display the result. Another to help teach function may generate random functions that the class (and teacher) try to identify by asking for the images of their chosen numbers. A program to help in the interpretation of time-graphs can allow the class to specify the time at which events occur and display the corresponding graph.

19.1.4 Generalised 'test-beds'

Certain programs may be used to illustrate and explore ideas in many different situations. Parameters may be varied one by one and many examples experienced in a short space of time. Graphical programs such as a function plotting program and programs for 2D and 3D geometric transformations allow the class and teacher to explore many situations from those of a quite elementary nature such as the behaviour of $y=mx$ to problems of discontinuity and asymptotic behaviour and from the non-commutativity of certain transformations to the nature of a three-dimensional shear.

19.1.5 Mathsworlds

Many programs have applications that are of great appeal and interest to some children – the output may be aesthetically pleasing as with a piece of design or it could be stimulating as with a game. It is possible to design the input structure of such programs to make mathematics the language of communication between the child and the machine. Thus a spaceship may be controlled by instructions such as 'Turn left 30°' rather than just by turning a knob the appropriate amount in a video-game. A child need not ever have met measurement of angle in degrees before ,– a few plays of the game will soon enable him to make a mental picture of an angle and he will acquire a feel for its scale of measurement as an 'on-the-job' skill. A very sophisticated development of this idea can be found in Seymour Papert's book *Mindstorms*. The LOGO language that Papert developed has now been implemented on many classroom microcomputers.

19.1.6 Classroom programming

Each of the previous applications has pre-supposed the existence of some prepared programs – perhaps written by the teacher or brought from a software library or educational publisher. However, nearly all the programming examples presented in this book could be developed in the classroom as part of the normal lesson – many useful programs can be written in a dozen or so lines that only require quite an elementary knowledge of programming. This could involve groups of pupils using several microcomputers but it is also quite feasible to undertake it as a class exercise with the teacher (or nominee) at the keyboard and a couple of large screens for the class to see. This catalogue of approaches is by no means exhaustive but it does show that there are many possible teaching styles that can be used with a microcomputer, each of which will have its own organisational and resource implications.

19.2 ORGANISATION

For individuals to use a microcomputer to catch up on some work that they missed, the computer might need to be set up at the back or side of the class so as not to distract the others from their normal work – only a small screen is needed and the whole equipment could sit on a trolley that may be wheeled in when needed. For a group of pupils to be able to use a program to explore some mathematics on their own or to test specific skills it is necessary to devise some rota of activities (common in primary classrooms) so that each group gets their fair share and is gainfully occupied when not using the computer. For the whole class to be involved with a particular program it is vital that they should all be able to see the screen output. Only one computer is needed but even a pair of large second-hand TVs may not give clear enough definition for all to see in their normal places. This may imply using programs in which all text is produced in oversize characters or it may mean re-organising the seating for these activities. For some problem-solving and investigative activities it may be beneficial to have a practical workshop layout in which several computers, calculators and other apparatus are available for pupils to use as appropriate or as directed.

19.3 RESOURCES

If a department is to become actively involved in using microcomputers in its mathematics teaching it must have a resourcing policy (which may involve fund-raising in addition to school allocations). Unlike the language-laboratory experience of some years ago it is quite possible to make a long-term plan and to move towards that by small acquisitions and physical changes each year, though some may be in the fortunate position of being able to buy a complete set of equipment all at one time. The piecemeal approach needs to be handled with care lest technical developments lead to one each of several incompatible models. Though breakdowns are fortunately not frequent, choosing the one system for subsequent machines does have advantages.

19.3.1 Hardware

The first thing to determine is whether it is feasible for each department to go its own way in computer developments (so-called 'distributed computing') or whether the school has a policy to centralise such resources. If the latter is the case then, short of changing the policy, all that remains to be determined is whether individual items of equipment can be booked to departments for specific lessons, (or for longer periods of time), or whether there is a special room, or rooms, in which the equipment is to be used, and, if so, whether that can be booked for department use. Obviously where Computer Studies is taught as an examination subject the availability of equipment for departmental use is likely to be more restricted. If it is possible to develop an independent (or co-ordinated) policy then, again, a decision has to be made whether to aim for one specially equipped room or to go for a deployable resource that could be moved into any appropriate classroom. Security may well be a consideration here. Equipment on trolleys may be locked away in storage cupboards when out of use, but it may well be easier to make a single room secure. Again it is of great importance to clarify the LEAs attitude to insurance. In particular it may well be more prudent, as well as good staff development, to let staff take equipment home in holidays and half-terms, but the insurance position needs to be carefully checked. A policy adopted by some schools is to aim for a departmental collection of microcomputers on trolleys kept in a central store but to equip a couple of classrooms with large (old!) televisions mounted permanently on wall brackets or on cupboards so that individual/group work could take place in any suitable area but that class teaching using the computer is confined to those specifically equipped rooms. With the rapid fall in prices and the ever-widening range of machines on the market it does now become a feasible proposition for a department to aim at acquiring, say, at least one microcomputer a year for the foreseeable future! It may need the help of parents, friendly firms and free-enterprise activities or it may need a case to be put for some change in the way money is allocated within the school, but with micros as cheap as £50 and with a wide range of machines for under £200 many schools could adopt such a policy. Of course it may be that the LEA has a 'one machine' policy and if that happens to cost, say, £1000 the whole picture is radically altered. There will, inevitably, be problems of compatibility of hardware and on no account should you rush into buying a machine just because it looks good on paper or has had a favourable review on some magazine. As you develop the usage of microcomputers one of the most valuable assets will be the stock of programs (perhaps bought from publishers or written by teachers, pupils or friends) and the collection of experience you have in the use of a particular model (or models) of microcomputer. Acquiring a totally different kind of micro may have considerable implications for the amount of time to be spent on transferring software and gaining experience in its use. The Council for Educational Technology (CET) publish some helpful specifications such as USPECs 32, 32a and 32b.

19.3.2 Materials

The most obvious material cost in the departmental use of micros is that of program storage. For programs that are to be frequently used you will need at least two, and probably more, copies kept separately. If you use cassettes then you can aim to keep several 'secure' copies together on longer tape (for example, C90) but, in general, it is inconvenient to have more than one 'working' copy of a program per side of a short cassette (for example, C12). You must, then, budget for quite a fair supply of tapes per year and you should check if there is a county supplies discount. The same considerations will apply to floppy discs and printer paper, where used. It is possible to use the back of previously used computer printout paper. It may be necessary to budget small amounts for simple electronic components, plugs, wires, solder etc., but on the whole it pays to be on good terms with the science or technical studies departments! It may also be prudent to budget a small contingency amount on repairs, but, again, with the right help repairs can be carried out for minimal sums.

19.3.3 Software

Many educational publishers are now becoming involved in the business of selling educational software packages. Sometimes these are the products of nationally funded development programmes such as those sponsored by the Schools Council and the Microelectronics Education Programme. The prices of individual programs and their associated documentation, pupil notes etc. vary widely and, of course, in view of the ease of copyright infringements by duplicating tapes and discs and photocopying documentation, publishers are seldom willing to offer materials on approval. Again it pays to have access to an adviser, Teachers' Centre, college or other schools in order to see the software before purchasing it. However, the evaluation of a piece of educational software is quite a complex and time-consuming task. You need to consider *technical* qualities, such as its ease of use, response to incorrect inputs, text layout etc., *educational* qualities, such as where it fits into the curriculum, what kind of pupils it might be appropriate for, whether its approach could be just as easily fulfilled without a computer etc., and *practical* qualities, such as what classroom organisation is required, whether it is suitably motivational for pupils, whether it is straightforward for colleagues to use etc.

Whatever the source of software it will be necessary to maintain a software library so that access to the programs and associated documentation and materials is convenient and secure. Responsible pupils can be a great help in maintaining such a library, and in routine jobs such as cleaning tape heads and producing back-up copies of software. Where home-produced software is used it is advisable to have some sort of departmental guidelines for authors to aid maintainance, portability and use. Thus it will be useful to have a model program available which illustrates whatever structuring and program standards are desired and which is documented to the required standard.

19.4 RANGE OF ABILITY

It may be that part of a departmental policy will consist of determination of priorities for use with particular age groups or ranges of abilities. Evidence is short but there are reports of the motivating effects of microcomputer use with both extremes of the ability range. A computer program may appear to have patience and to be even-handed and uncritical but the computer itself is an incomparable resource for the posing of mind-stretching problems. There is a strong case for the acquisition of programs suitable for the lower ability pupils and for the development of a set of problems and projects to stretch the abilities of the ablest pupils.

19.5 DEPARTMENTAL ACTIVITIES

In addition to the use of microcomputers within the main stream of mathematics teaching there are other activities that a department may wish to consider. One often provided activity is the informal lunchtime and after-school use of the computers by pupils 'doing their own thing' – it may that a rigid booking system is needed and that particular encouragement should be given to girls and to more reticent pupils to become involved, as they might otherwise be crowded-out by the inevitable corps of enthusiasts.

Attention may also be given to the need to provide time for staff use – both for 'in-service' activities and for program development. A colleague who can only get his hands on a machine for an hour or so a week is not likely to get very far in developing his use of the microcomputer – there really is no substitute for letting him have extended use of a machine at home over a weekend or in the holidays. Another matter to be considered is whether the computer has a role in departmental administration. An obvious application is in maintaining form and set lists with the use of a printer. There are also commercially produced programs to help in allocating pupil option choices and, to some extent, with timetabling – though most of these do require use of a disc drive. Perhaps not so obvious is the application of a word-processing program to the production of worksheets, handouts and examination papers. Word-processing consists essentially of two activities: editing and formatting. Text can be typed in, or loaded in from disc or tape and then changed as desired on the screen. Words or lines can be removed or replaced, whole paragraphs can be shifted round and so on. The edited copy can then be saved onto cassette or disc if required. Formatting allows you to specify the way it is to be printed out, for example, single or double spaced, left or right justified (or both), number of lines per page, number of characters per line etc. The usual kind of microcomputer printer is a 'dot-matrix' printer where each character is formed by the firing of a particular pattern of needles at the ink-filled ribbon and its printing rate is typically around 40 characters per second. Thus a fairly full A4 sheet of notes may take around 2 minutes to print. The printer could be used for mass-production of text or it could be used to generate masters from which copies are made by photocopying, off-setting or spirit duplicating. To prepare a spirit duplicator master it is often possible to remove the printer's ribbon and to allow the needles to strike the master directly, thus giving a better impression from the ink-filled backing sheet. For higher quality printing a 'daisy-wheel' printer or 'spin-writer' is needed, but these cost considerably more (around £500–£1000 at time of writing). However, there are now a range of daisy-wheel electronic typewriters that can be fitted with a computer interface (parallel or serial) and which have the advantage that they can earn their keep as a typewriter for normal use and just be hooked up to the computer when top-quality printing is required.

19.6 STAFF

One very important question of attitude to microcomputers is fundamental to any policy of staff development – 'Do mathematics teachers need to be able to program?' This question needs a little further clarification since 'programming' might be taken to mean the production of intricate educational programs. In this context, by programming we mean the activity of preparing a computer program to solve a mathematical problem – a problem-solving skill to put alongside, say, the use of a calculator. Only comparatively few teachers need to be able to produce educational programs – and those that aim to do so should take careful stock of the time involved. But is it sufficient for the others just to be trained to use the computer as a piece of educational technology, like a video recorder, so that they may be able to use pre-pared programs? The trouble is that it takes comparatively little effort to learn enough of, say, the BASIC programming lanaguage to understand the programs presented in this book but that the associated problem-solving skills that enable a problem to be transformed into a computer program are far from easily acquired. The issue, then, goes much further than use of microcomputers and has much to do with the feasibility of implementing the Cockcroft report's recommendation §243 which includes opportunities for problem-solving as an essential feature of mathematics teaching at all levels. Thus, in the short term, it may be feasible to train all members of a department to use the microcomputer, but the task of instilling confidence in them to use the micro as a problem-solving tool is much trickier and may well take a long time. A good self-study pack, such as that published for teachers by the Open University, may be a very valuable resource. There may, then, be a need for training sessions, or courses, in the use of micro and in, perhaps, the evaluation of educational software, but such a structure might not be appropriate for developing problem-solving skills which may need a much more informal co-operative workshop. With any policy of staff development it may well be wise to consult departments in neighbouring schools to exchange views on these issues and, perhaps, to arrange cooperative activities where feasible. Apart from issues of staff development a policy may need to be adopted with regard to time and effort expended on program development and on other areas of the school curriculum where computers may be involved. An obvious issue arises with the teaching of Computer Studies – should it be seen as the task of a mathematics department to assist in, or even to take responsibility, for such courses? If so, what else suffers and, if not, where else is the expertise to be found in the school? In many cases mathematics staff have been involved in such courses, but there are now many in-service courses for potential teachers of the subject. A school policy could be taken which (if adopted) would free mathematics teachers from this task in the medium term. Program development can be extremely time-consuming and, to some extent, addictive (golf-widows are being outnumbered by computer-widows!) so those involved in it will need to keep a close check to see that their other professional responsibilities are not suffering. The issues referred to in the earlier discussion on staff development also have a serious bearing on the content of initial teacher training of mathematics teachers. It may be necessary to say that all teachers in training should become familiar with educational applications of microcomputers and to have gained some experience in their use. But that may not be sufficient for intending mathematics teachers – should they, too, be able to program and, if so how can this be achieved and assessed?

19.7 HELP

An almost essential requirement for a department embarking on the use of microcomputers is a good network of contacts who can be approached for help and advice. To take a simple example, we have an expensive microcomputer that has worked reliably for a couple of years. One day on switching on, it produced a weird pattern of stripes on the screen and a nasty buzzing sound. What steps would you now take? Phoning a service company to send an engineer out will inevitably result in a large bill – and possibly the loss of the machine for a period. Taking it to a service company may save expense but takes time and does not necessarily guarantee any greater chance of success. However, the phone number of a contact familiar with that machine can (and did) save much time and money. In this example all that was required was to lift the lid and carefully (with the machine switched off!) press down all the chips in turn so that they were securely seated in their sockets – the only fault was a bad contact. In another microcomputer we have learnt that incorrect insertion of the tape-cassette lead into the computer may cause a particular chip to 'blow'. This chip costs about £4 and we keep a few in store in case of such trouble. Apart from hardware advice you need to identify sources of information on software so that you can find out what is available, and, if possible, go and see it for yourself. Similarly it is very heplful to know of other schools (not necessarily particularly local) that are making use of microcomputers so that you can pick up (or exchange) ideas. Again help may be needed within school with technical jobs such as soldering, making stands etc. and with routine jobs such as filing, cataloging, labelling etc. – do not underestimate the use of pupils as a resource, many of them will be far more familiar with aspects of microcomputers than you and your colleagues!

19.8 CURRICULUM

Reference has already been made to the Cockcroft recommendation §243 and there can be little doubt that a microcomputer is a most powerful tool for investigative and problem-solving activities in mathematics. It seems to make the question 'WHAT IF ...?' occur naturally and frequently and to provide the means of production of evidence to answer that very question. As we have tried to emphasise, there may be situations in which mathematical hypothesis-forming is inhibited by the sheer drudgery of data collection and where the use of microcomputer can relieve us of that drudgery to engage in the kind of mathematical discussion which also forms part of the recommendation. The major question to be considered is whether *programming* (in the sense we have been using it) should be part of the mathematics curriculum and if so at what stage should it be introduced and to which groups of pupils. Coupled with this is the question of assessment – since problem-solving skills (such as programming) are renowned for their incompatibility with standard methods of assessment and examination. Moreover, apart from the relationships between the use of microcomputers and the present mathematics curriculm, we need to keep the content of the curriculum under review and to determine whether it may need altering in the light of possible widespread use of computing and calculating equipment. It has been widely reported that the mathematics curriculum has hardly shown any perceivable change despite the widespread availability of cheap electronic calculators – will it remain equally impervious to the computer era or, if not, from what quarter will come the pressure to change?

Bibliography

[1] Abelson, Harold and DiSessa, Andrea. *Turtle Geometry: The Computer as a Medium for Exploring Mathematics*. MIT, 1981.

[2] Atherton, Roy. *Structured Progamming with COMAL.* Ellis Horwood, 1982.

[3] Bezier, P. E. *Numerical Control: Mathematics and Applications.* Wiley, 1972.

[4] Chasen, Sylvan H. *Geometric Principles and Procedures for Computer Graphic Applications.* Prentice-Hall, 1978.

[5] 'Chronicler, The'. *Fred Learns about Computers.* Continua, 1978.

[6] Cope, Tonia. *Computing using BASIC: An Interactive Approach.* Ellis Horwood, 1981.

[7] Engel, Arthur. *Elementarmathematik vom algorithmischen Standpunkt.* Ernest Klett, 1977.

[8] Faux, I. D. and Pratt, M. J. *Computational Geometry for Designing and Manufacture*. Ellis Horwood, 1980.

[9] Fletcher, Trevor J. *Linear Algebra through its applictions.* Van Nostrand Reinhold, 1972.

[10] *Great Britain Mathematics Counts: report of the Committee of Enquiry into the Teaching of Mathematics in Schools* under the chairmanship of W. H. Cockcroft. HMSO, 1982.

[11] Hampshire, Nick. *The PET Revealed*, 2nd edn. Computabits, 1980.

[12] Hofstadter, Douglas R. *Godel, Escher, Bach: An Eternal Golden Braid.* Penguin, 1980.

[13] Kemeny, John G. and Kurtz, Thomas E. *BASIC Programming*, 3rd edn. Wiley, 1980.

[14] Knuth, Donald E. *Art of Computer Programming*, Vol. 1: *Fundamental Algorithms*, Vol. 2: *Seminumerical Algorithms*. Addison Wesley, 1973 and 1981.

[15] Lighthill, James (Ed.). *Newer Uses of Mathematics.* Penguin, 1978.

[16] Newmann, William and Sproull, Robert. *Principles of Interactive Computer Graphics.* McGraw-Hill, 1973.

[17] Papert, Seymour. *Mindstorms: Children, Computers, Powerful Ideas.* Harvester, 1980.

[18] Rogers, David F. and Adams, James A. *Mathematical Elements of Computer Graphics.* McGraw-Hill, 1976.

[19] Smith I. C. H. (Ed.). *Microcomputers in Education.* Ellis Horwood, 1982.

[20] Tocher, Keith D. *The Art of Simulation.* EUP, 1964.

[21] Warwick, John (Ed.). *Working Notes on Microcomputers in Mathematical Education.* ATM, 1982.

[22] West, Raeto. *Progamming the PET/CBM.* 1982.

Articles and materials

There are two major groups for teachers concerned with the use of computers in schools, both of which publish regular journals which often contain articles of mathematical interest. They are (i) the Computer Education Group (CEG) which publishes *Computer Education* and (ii) Microcomputer Users in Secondary and Primary Education (MUSE) which produces *Computers in Schools.* There is also a commercially produced monthly journal: *Educational Computing.*

For each kind of microcomputer there is usually at least one User Group which publishes a regular newsletter or journal and, of course, newsagents stock many monthly magazines about computing. The Times Educational Supplement regularly produces supplements about mathematics and computers.

Programs for use in mathematics teaching are published form a variety of sources. Nearly all the major educational publishers are selling software which comes from many backgrounds (some from individual teachers others from major projects such as ITMA from Plymouth, Fiveways from Birmingham and the Schools Council Computers in the Curriculum project). The Microelectronics Education Program (MEP) has produced training packs for teachers and is commissioning development work in mathematics. The Association of Teachers of Mathematics has constituted a working group on microcomputers and is publishing programs and materials and further useful material has come from the Shell Centre at Nottingham University and from the Open University.

Most of the journals about mathematical education, such as *Teaching Mathematics and its Applications*, *Teaching Statistics*, *Mathematics in Schools*, the *Mathematical Gazette* and *Mathematics Teaching*, carry relevant articles from time to time and some of the articles that we have found useful are listed (in date order) below.

The Mathematical Gazzette

Vol. 65, No. 433: McGregor, J. J. and Watt, A. H. PASCAL rules O.K. Oct. 1981.

No. 432: Oldknow, A. J. An analysis of an algorithm. June 1981.

Vol. 63, No. 426: Manning, J. R. Probability distributions and B-splines. Dec. 1979.

Mathematics Teaching

No. 98: Hart, M. Using computers to understand mathematics, four years on. Mar. 1982.

No. 97: Engel, A. Numerical algorithms. Dec. 1981.

No. 94: Ahmed, A. G. and Oldknow, A. J. Scampi and chips – computers on the mathematics menu.

No. 93: Fraser, R. JANE. Dec. 1980.

No. 91: Oldknow, A. J. Some Notes from a microcomputer workshop. June 1980.

No. 80: Hart, M. Using computers to understand mathematics. Sep. 1977.

No. 72: Brissenden, T. F. and Davies, A. J. Computer graphics in the teaching of science and mathematics. Sep. 1975.

No. 67: Oldknow, A. J. Motion geometry in action. June 1974.

No. 58: Berryman, J. The mathematics of mathematical notation. Apr. 1972.

No. 58: Papert, S. Teaching children thinking. Apr. 1972.

No. 57: Berryman, J. Computer science – a current threat. Dec. 1971.

List of programs

Index